# Plumbing Licensing Study Guide

# Plumbing Licensing Study Guide

Rex Miller and
Mark R. Miller

Industrial Press

# Industrial Press, Inc.

32 Haviland Street, Suite 3
South Norwalk, Connecticut 06854
Phone: 203-956-5593
Toll-Free in USA: 888-528-7852
Fax: 203-354-9391
Email: info@industrialpress.com

Author: Rex Miller and Mark Miller
Title: Plumbing Licensing Study Guide
Library of Congress Control Number: 2018932682

© by Industrial Press, Inc.
All rights reserved. Published in 2018.
Printed in the United States of America.

ISBN (print): 978-0-8311-3625-3
ISBN (ePUB): 978-0-8311-9471-0
ISBN (eMOBI): 978-0-8311-9472-7
ISBN (ePDF): 978-0-8311-9470-3

Editorial Director: Judy Bass
Copy Editor: Janice Gold
Compositor: Patricia Wallenburg, TypeWriting
Cover Designer: Janet Romano-Murray

industrialpress.com
ebooks.industrialpress.com

# Contents

## 19  Measurement Systems (U.S. to Metric)  149

## 20  Measuring Weights and Liquids  159

## A  Techniques for Studying and Test-Taking   225

## B  Practice Questions on Plumbing   233

## C  Answers to Test Questions   247

## D  Plumbing Specifications   255

## Index   259

# Preface

This edition of *Plumbing Licensing Study Guide* is just as its name implies—a study guide. It is designed to be used as a text, a reference source, and to aid the reader in preparing for the examination that is important in becoming a professional. Students and others interested in the plumbing trade at all levels will find the book both informative and interesting.

The value of this edition has been enhanced by the information contained therein. The information is presented for ease in studying the essentials a plumber must possess when working in the trade.

The main purpose of this book is to aid you in your everyday tasks and keep you updated with the latest facts, figures, and devices in this important trade.

Many illustrations are included, which show a variety of parts and techniques found in present day practice in the field. Obviously, not all related problems can be presented here, since there is a great deal of ingenuity required by the worker on the job. For standard procedure, however, the various codes do give a guide to the size, type and kind of pipe allowed in the many locations found in today's construction. This book should be part of your tool kit. Many tables are included to aid you in quickly being the source for an answer.

# Acknowledgments

The authors would like to thank the following manufacturers, institutes, and others for their generous efforts. They furnished photographs, drawings, and technical assistance. Without their valuable time and effort this book would not have been possible. We hope this acknowledgment of some of the contributors will let you know that the field you are working in, or are about to enter, is one of the best. Individuals, too numerous to mention, have also played a role in this book. We would like to take this opportunity to thank them for their contributions.

- ABS Pumps
- American National Standards Institute (ANSI)
- Charlotte Pipe Company
- CISPI
- Cast Iron Soil Pipe Institute

- Michael Steele Crane Company
- Electric Eel Manufacturing Co.
- Eljer Plumbingware Co.
- EPA (Environmental Protection Agency)
- Fernco, Inc.
- Flexcon Industries, Inc.
- Genova, Inc.
- Gould Pumps, Inc.
- Jasam Manufacturing Co.
- National Capital Poison Center
- Nibco Co.
- Plueger Submersible Pump Co., Inc.
- Plumb Shop
- Plumbing Manufacturers Institute
- Ridge Tool Co.
- Sloan Valve Co.
- Teflon
- Wrightway Manufacturing Co.

# Plumbing Licensing Study Guide

# 1 | Tools

## Performance Objectives

After studying this chapter, you will be able to:

- Identify the tools a plumber uses in his daily work.

- Explain how the various hand tools operate and where.

- Operate various electronic machines that locate buried pipes, power lines, and telephone and cable lines.

- Be able to answer the review questions at the end of the chapter.

1

# Plumber's Tools

Tools are the major expense in setting up a plumbing operation—private or public. To make a system operational, it is necessary to have the proper tools. The choice of tools is dictated by the type of job. There are catalogs full of devices labeled as tools that eliminate involved procedures or save you time or money while on the job. Shown here are a selected few of the more mundane tools or those used on every job and some specialty items that may have to be acquired to properly get a job done in an efficient and timely manner.

One of the most often used sources of information on tools is the Internet. For example,

Ridge Tool Company has a website with all of its products and they provide an owner's manual in PDF format for downloading. Most of the tools are represented with detailed instructions on how to safely use them. A few of the more often used and special items are shown here. For more details just put Ridge Tools in the search blank and hit search. Open the first or number one item on the list for access to the catalog with available Owner's Manuals and Parts List. Some of the tools to be found in catalogs of manufacturers are shown here in various figures. These are listed to help you in learning to identify and easily recognize the tool and its use.

**Figure 1-1**    Ball peen hammer. (Ridge Tool Co.)

**Figure 1-2**    Pipe wrench. (Ridge Tool Co.)

**Figure 1-3**    Cast iron cutter. (Ridge Tool Co.)

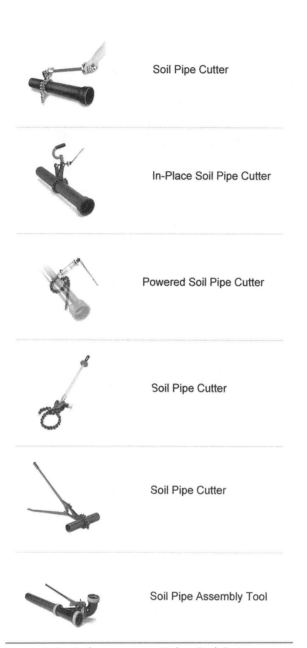

Soil Pipe Cutter

In-Place Soil Pipe Cutter

Powered Soil Pipe Cutter

Soil Pipe Cutter

Soil Pipe Cutter

Soil Pipe Assembly Tool

**Figure 1-4**  Soil pipe cutter. (Ridge Tool Co.)

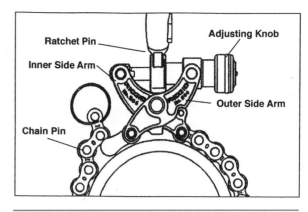

**Figure 1-5**  Close-up view of chain cutter. (Ridge Tool Co.)

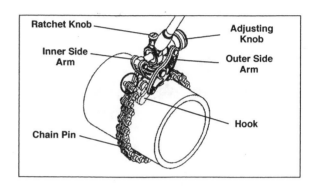

**Figure 1-6**  Closer look at the soil pipe cutter. (Ridge Tool Co.)

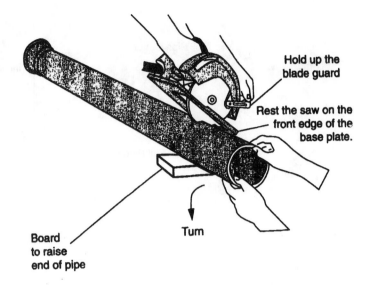

Hold up the
blade guard

Rest the saw on the
front edge of the
base plate.

Turn

Board
to raise
end of pipe

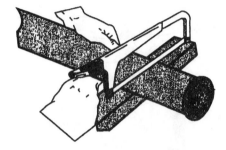

1. Use a hacksaw to
make a $^1/_{16}$"-deep cut
around the pipe.

2. Then deepen the cut
with a cold chisel.

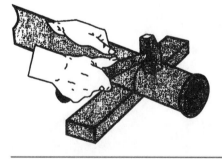

3. Tap the end all the way around.

**Figure 1-7**   Soil pipe cutter. (Ridge Tool Co.)

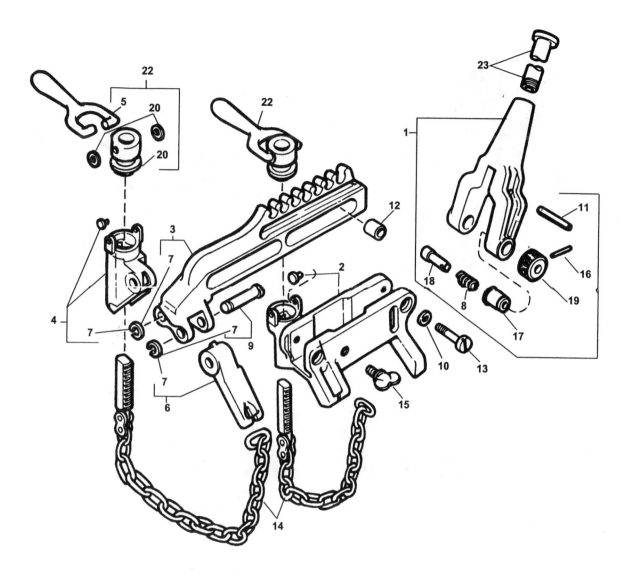

**Figure 1-8** Soil pipe assembly tool with reference numbers. (Ridge Tool Co.)

| Ref. No. | Description | Ref. No. | Description |
|---|---|---|---|
| — | 228 Soil Pipe Assembly Tool | 12 | Slide Roller (2) |
| 1 | Yoke Assembly | 13 | Slide Roller Screw (2) |
| 2 | Housing w/Pivot Pin | 14 | Chain and Screw Assembly (2) |
| 3 | Slide Assembly | 15 | Thumb Screw |
| 4 | Hinged Jaw Assembly | 16 | Pin (2) |
| 5 | Handle (2) | 17 | Insert (2) |
| 6 | Hinged Hook Assembly | 18 | Pivot Pin (2) |
| 7 | Retaining Ring (2) | 19 | Knob (2) |
| 8 | Coil Spring (2) | 20 | Washer (4) |
| 9 | Hinge Pin Assembly (2) | 21 | Swivel Nut (2) |
| 10 | Lock Washer (2) | 22 | Handle and Nut Assembly (2) |
| 11 | Ratchet Pin | 23 | Handle |

**Figure 1-9**   Swaging tool set. (Ridge Tool Co.)

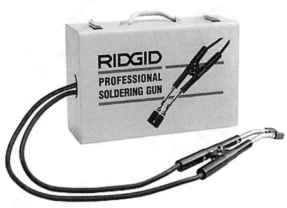

**Figure 1-10**   Soldering gun. (Ridge Tool Co.)

**Figure 1-11**   Pipe thawing unit. (Ridge Tool Co.)

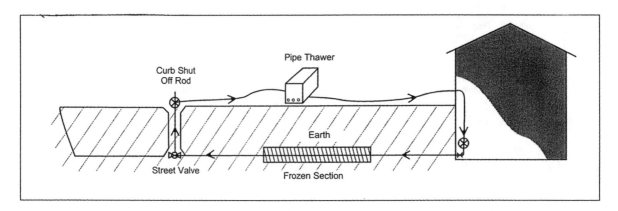

**Figure 1-12**   Thawing main service lines. (Ridge Tool Co.)

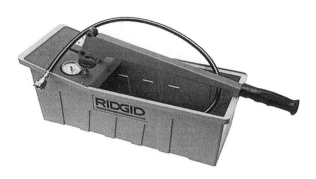

**Figure 1-13** Pressure tester. (Ridge Tool Co.)

**Figure 1-16** Heavy duty pipe cutter—cuts ½-inch through 2 inches (3 through 51 mm). (Ridge Tool Co.)

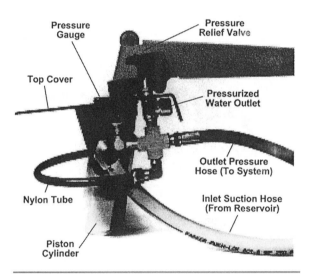

Pressure Gauge
Pressure Relief Valve
Top Cover
Pressurized Water Outlet
Outlet Pressure Hose (To System)
Nylon Tube
Inlet Suction Hose (From Reservoir)
Piston Cylinder

**Figure 1-14** Details of components of the Model 1450 pressure tester. (Ridge Tool Co.)

**Figure 1-17** Channel-lock pliers. (Ridge Tool Co.)

**Figure 1-15** Straight pipe wrench. (Ridge Tool Co.)

**Figure 1-18** Pipe and bolt threading machine for pipe ½-inch through 2 inches (3 through 51 mm). The machine cuts, threads, and oils. (Ridge Tool Co.)

**Figure 1-19**  Adjustable wrench also known as a crescent wrench. (Ridge Tool Co.)

**Figure 1-20**  Internal wrench. This wrench is used to hold closet spuds and bath, basin, and sink strainers through 2 inches (51 mm). Also used for installing or extracting 1-inch through 2-inch (25 mm through 41 mm) nipples without damage to the threads. (Ridge Tool Co.)

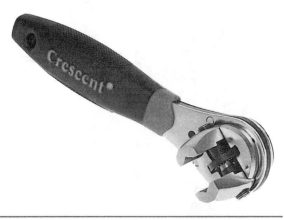

**Figure 1-21**  Basin wrenches.

**Figure 1-22**  Spud wrench.

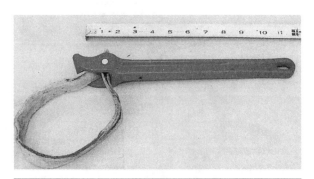

**Figure 1-23**  Strap wrench.

**Figure 1-24**  Quick-acting tubing cutter for ¼-inch through 2⅝-inch tubing (6 mm thru 66.5 mm). (Ridge Tool Co.)

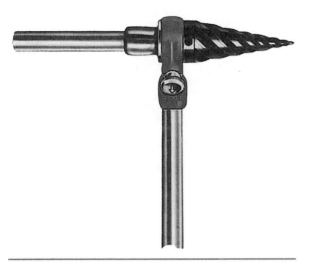

**Figure 1-25**   Spiral ratchet pipe reamer, ½-inch
through 2 inches (3 mm through 51 mm).
(Ridge Tool Co.)

# Review Questions

1. How do you select a tool for your toolbox?

   _____

2. How many types of pliers do you need for the toolbox?

   _____

3. Describe the internal wrench.

   _____

4. Where would you need a crescent wrench?

   _____

5. What is the word nipple used to describe in plumbing jargon?

   _____

6. What is a soil pipe made of?

   _____

7. Where are soil pipes used?

   _____

8. Where would you use a spud wrench?

   _____

# 2 | Plumbing Terms

## Performance Objectives

After studying this chapter, you will:

- Understand the plumbing trade's terms and tools.

- Know how brass, copper, and other metals are used in plumbing work.

- Know the national standards that were created and why they are needed.

- Be able to identify plumbing tools and how they are used on the job.

- Know why leaching of metals is important to prevent.

- Be able to describe how a pressure valve works.

- Know how ADA affects plumbers.

- Be able to answer the review questions at the end of the chapter.

# Plumbing Industry

- **ANSI.** The American National Standards Institute (ANSI) is a not-for-profit, non-government organization that oversees the creation and use of voluntary health and safety standards for products and businesses across nearly all sectors of the U.S. economy.
- **Brass.** An ancient alloy, composed primarily of copper and zinc, used in the manufacture of faucets and other plumbing fittings. Small amounts of other alloying materials are also added for various types of brass to address the requirements of specific applications. Brass is also the term for a faucet finish, also known as polished brass.
- **Copper.** One of the basic elements (Cu), copper is used for plumbing piping, and is one of the alloys used to make brass, a key material in the manufacture of faucets and fittings.
- **Fitting.** A device designed to control and guide the flow of water. Examples include faucets, shower heads, shutoff valves, shower valves, and drinking fountain spouts. Some people call these fixtures, but that term means something different to the plumbing industry. The differing usage of "fittings vs. fixtures" can lead to unintended consequences, such as when legislation calls for changes in fixtures, although the true intent involves changes in fittings (see "Fixture").
- **Fixture.** A device for receiving water and/or waste matter that directs these substances into a sanitary drainage system. Examples include toilets, sinks, bathtubs, shower receptors, and water closet bowls. The term is used erroneously in common vernacular to describe fittings (see "Fitting").
- **Gallons per minute** (GPM or gpm). A measure of the rate at which water flows through a fixture or fitting at a certain pressure. It is measured by the number of gallons flowing from the device in one minute at a given water supply pressure.
- **Gallons per flush** (GPF or gpf). A measure of the total volume of water required to flush a water closet or urinal, measured in gallons.
- **Lavatory.** While sometimes used by the general public to mean a bathroom or washroom, the plumbing industry uses lavatory to mean a bathroom washbowl or basin permanently installed with running water. The plumbing industry uses the term "sink" in reference to kitchen sinks.
- **Performance-based product standards.** These standards define a desired outcome from products, related to what they do, rather than how they are made and what they are made of. These standards typically prescribe a means for determining whether the product delivers to the standard. For example, a performance-based product standard; for fittings would specify the maximum amount of alloy materials, such as lead, that may be leached into the drinking water.
- **Porcelain enamel.** A coating used on metal fixtures, such as cast iron sinks and bathtubs. Ceramic material is fired at high temperature to form a vitreous porcelain film that is fused to the base metal of the fixture or fused to a ground coat. Porcelain enamel gives metal plumbing fixtures their colors and desirable glossy surfaces.
- **Plumbing Manufacturers Institute (PMI).** A not-for-profit trade association of plumbing products manufacturers. PMI member companies produce most of the nation's plumbing products.
- **Product standards.** Established by research and consensus, product standards define what products are made of, as well as how they perform. For plumbing products, product standards govern the characteristics,

materials, performance and operability, as well as how products need to interact with other plumbing-system elements. For example, a product standard for fittings would define the alloys and the amounts that can be used in their manufacture.

- **Prescriptive product standards.** These standards differ from performance-based product standards in that they attempt to achieve a desired outcome by specifying the characteristics, materials, performance and operability of products. For example, a prescriptive product standard for fittings would specify the maximum amount of alloy material, such as lead, that can come in direct contact with the drinking water.

- **PVD finishes.** PVD stands for physical vapor deposition. This process, which occurs in a vacuum chamber, electro-statically applies extremely thin, but extremely dense coatings of exotic metal alloys onto fittings. The resulting finish is state-of-the-art in durability, scratch resistance and lasting beauty for faucets. A wide range of finishes with PVD is possible, including chrome, nickel, brass and bronze.

- **Valve.** A fitting with a movable part that opens or closes one or more passages and thereby allows a liquid flow to be started, stopped, and regulated. In plumbing, valves are used in faucets and showers, and can be called mixing valves because they control the mix of hot and cold water to achieve desired water temperatures.

- **Vitreous china.** A type of pottery most commonly used for plumbing fixtures, such as toilets. It is a compound of ceramic materials fired at a high temperature to form a nonporous body. Exposed surfaces are coated with a ceramic glaze that fuses to the china when fired and gives vitreous china plumbing fixtures their colors and glossy appearance.

# Clean Water

- **Leach.** In the case of plumbing systems, leaching refers to the process of dissolving a soluble component out of a constituent material at a wetted surface. Materials commonly leached into drinking water from water distribution systems include copper, lead, and nickel.

- **Lead.** One of the basic elements (Pb), lead is a soft metal that has been used in plumbing systems for thousands of years. The word "plumbing" derives from the Latin word for lead, plumbium. Lead has a unique ability to resist pinhole leaks, while being soft enough to form into shapes that deliver water most efficiently. Its softness and malleability were for a long time highly desirable properties for manufacturing everything from pipe to paint. Lead is a neurotoxin that can accumulate in the body in soft tissues, as well as bone.

- **Lead and Copper Rule (LCR).** A United States Environmental Protection Agency regulation dating back to 1991, LCR requires water systems to monitor drinking water that comes through faucets in homes and buildings. If lead concentrations exceed 15 parts per billion (ppb) or copper concentrations exceed 1.3 parts per million (ppm) in more than 10 percent of homes and businesses sampled in a regional plumbing system, the system must take actions to control corrosion and leaching. If the action level for lead is exceeded, the system must also inform the public about steps they should take to protect their health, including the possible replacement of plumbing system piping.

- **Lead-free.** Under section 1417(d) of the Safe Drinking Water Act, "lead free" is defined as being no more than 0.2 percent of materials used in solders, and no more than 8 percent of materials used to manufacture pipe, fittings, and well pumps.

- **National Primary Drinking Water Regulations** (NPDWRs, or primary standards). Legally enforceable federal standards that apply to public water systems. Primary standards limit the levels of contaminants in drinking water. A 1996 amendment to the Safe Drinking Water Act (SDWA) requires that the United States Environmental Protection Agency establish a list of contaminants every five years that are known or anticipated to occur in public water systems and may require future regulations under the SDWA.

- **National Science Foundation** (NSF). Founded in 1944, NSF International is a not-for-profit, non-governmental organization that develops standards and product certifications in the area of public health and safety.

- **NSF/ANSI Standard 60.** A standard related to chemicals used to treat drinking water. Developed by NSF and conforming to the ANSI voluntary standard, the standard was accepted by the NSF board in 1988 to evaluate products, such as softeners and oxidizers, to assure that usage amounts safeguard the public health and safety.

- **NSF/ANSI Standard 61.** A standard related to products that come in contact with drinking water. Developed by NSF and conforming to the ANSI voluntary standard, the standard was accepted by the NSF board in 1988 to confirm that such products will not contribute excessive levels of contaminants into drinking water. Most U.S. states and many Canadian provinces require products used in municipal water distribution systems and building plumbing systems to comply with Standard 61.

- **Potable water.** Water that is satisfactory for drinking, culinary and domestic purposes.

- **Proposition 65.** Also known colloquially as Prop 65, California's Safe Drinking Water and Toxic Enforcement Act of 1986 requires companies to post notice of chemicals in products that can be released into the environment and have been determined by the state to be a cause of cancer. In early 2008, the list included 775 chemicals. Prop 65 impacts residents in other states when they receive such notices in purchased products, such as bathroom faucets. Companies will often post the notification on all products, rather than incur extra costs to isolate products sold only in California.

- **Safe Drinking Water Act** (SDWA). SDWA is a federal law originally passed by Congress in 1974 to protect public health by regulating the nation's public drinking water supply. Amendments were passed in 1986 and in 1996. The SDWA requires many actions to protect drinking water and its sources: rivers, lakes, reservoirs, springs, and ground water wells. SDWA authorizes the United States Environmental Protection Agency to set national health-based standards for drinking water to protect against both naturally occurring and man-made contaminants. Enforcement is accomplished through the National Primary Drinking Water Regulations.

## Water Efficiency

- **Dual-flush.** A high-efficiency toilet that gives users the choice of flushing with the maximum amount of water allowed by law (1.6 gpf in the United States) or less water. The average amount of water used by the toilet cannot be more than 20 percent less than the maximum allowable, qualifying it to be considered high-efficiency and eligible for Water Sense labeling.

- **Energy Policy Act of 1992.** Among the provisions of this federal legislation, the Energy Policy Act of 1992 required that all residential toilets had to flush using no more than 1.6 gallons per flush.

- **Flapper.** The moveable part of a toilet flush valve that releases the water from the tank into the bowl when the toilet is flushed and seals the valve shut when the toilet is not being flushed. The most familiar version is a red replacement rubber ring that can deteriorate over time, leading to leaks and water waste. Newer flapper technologies are impervious to deterioration, increasing water efficiency and reducing operating costs.

- **Flush valve.** Located at the bottom of a toilet tank, the flush valve discharges the water from the tank into the bowl when the toilet is flushed.

- **Gravity-fed toilets.** The most common type of toilet in the United States, gravity-fed toilets rely on the force of gravity to flush the toilet effectively. The natural force of water dropping down from the tank scours the bowl clean and forces water and waste quickly into the trap-way.

- **High efficiency toilet** (HET). A toilet with an average water consumption of 1.28 gallons per flush or 4.8 liters per flush, when tested in accordance with a standard or product specification, such as the United States Environmental Protection Agency's Water Sense program. HETs use 20 percent less water than mandated by the Energy Policy Act of 1992, which lowers utility bills and reduces the strain on septic systems. HETs are eligible for special rebates in many drought prone areas. They are available as single flush gravity toilets, dual flush gravity toilets or pressure-assisted toilets.

- **High efficiency urinal** (HEU). A urinal that uses a half gallon or less of water, half the amount allowed under the Energy Policy Act of 1992. It contributes to lower utility bills, while reducing the burden on septic systems. HEUs are sometimes thought to be waterless, which isn't true. Waterless urinals are one type of HEU, but there are also urinals that use water and still meet higher efficiency standards.

- **Low-flow.** In the plumbing industry, low-flow fixtures and fittings refer to plumbing products that meet the water efficiency standard of the Energy Policy Act of 1992. The term is used interchangeably with the term "low consumption."

- **MaP testing.** A voluntary test protocol for toilets that measures the ability to remove solid waste, also referred to as "bulk." Cooperatively developed in 2003 by water utilities and water efficiency specialists in the United States and Canada, it uses soybean paste (miso) as test media, in an effort to replicate "real world" waste. The test is conducted by successively increasing the amount of test media that is flushed until the toilet is no longer able to reliably or completely remove the media from the bowl. Results are reported as a MaP score, which is related to the number of grams of a test media that a toilet can adequately flush.

- **Metered toilets or metered flush.** A toilet with a mechanism that delivers a precise, non-variable amount of water with each flush.

- **Pressure-assisted toilets.** A toilet that uses a compressed-air device to enhance the force of gravity used to clean the bowl when the toilet is flushed.

- **Trap-way.** The channel in a toilet that connects the bowl to the waste outlet. The trap-way is measured in terms of the largest diameter ball that can pass through it, called a ball-pass or ball passage.

- **Ultra-low-flow.** In the plumbing industry, ultra-low flow fixtures and fittings refer to plumbing products that exceed the water efficiency standard of the Energy Policy Act of 1992. The term is used interchangeably with the term "high efficiency."

■ **Water Sense.** Water Sense is a partnership program sponsored by United States Environmental Protection Agency, which works to promote water efficiency and enhance the market for water efficient products, programs, and practices. Similar to the Energy Star program that helps consumers choose energy-efficient appliances, Water Sense helps consumers to choose water efficient products by specifying the maximum flow rates and minimum performance levels. Products certified as meeting current Water Sense product specifications are eligible to carry the Water Sense label.

# Health and Safety

■ **Accessible design.** An approach to designing buildings, homes and products that renders them easier to access and use by people with physical, sensory, or cognitive disabilities.

■ **ADA-compliant device.** A device which is fully compliant, when properly installed, with the current requirements of the Americans with Disabilities Act Accessibility Guidelines (ADAAG), as legislated by the Americans with Disabilities Act of 1990.

■ **Americans with Disabilities Act** (ADA). A federal law, passed in 1990, which prohibits discrimination against people with disabilities. The term "disability" means a physical or mental impairment that substantially limits one or more of the major life activities of such individuals. Among the provisions in the law are requirements that impact plumbing products in the design of accessible bathrooms and facilities.

■ **Automatic compensating valve.** A valve that is supplied with hot and cold water and provides a means of automatically maintaining

the water temperature selected for an outlet. Automatic compensating valves are used to reduce the risk of scalding and thermal shock.

■ **Backflow.** A flowing back or reversal of the normal direction of wastewater from homes and buildings, leading to the possible contamination of potable water systems.

■ **Backflow prevention device.** Any mechanical device designed to automatically prevent backflow.

■ **Barrier-free.** Products and buildings are considered "barrier-free" if they permit access by all users, including those in wheelchairs. In plumbing products, the term can refer to showers that do not have a lip preventing wheelchair access, as well as sinks and water fountains that are usable at different heights.

■ **Pressure-balancing valve.** Also known as a pressure-compensating valve, this device is designed to reduce the risk of thermal shock and scalding while showering. Required by code in most areas of the United States, a pressure-balance valve senses the hot and cold water pressures coming in from the supply line and compensates for variations to maintain the water temperature. Such variations can occur when a toilet is flushed or a washing machine started while someone is showering.

■ **Proximity valves.** An electronic valve for plumbing fixtures and fittings that enables them to be operated without being touched. Similar to auto-open doors and light sensors that are activated by movement, proximity valves deliver the benefits of being both barrier-free and sanitary to use. Proximity valves can operate toilets, urinals and faucets.

■ **Thermostatic valves.** Also known as a thermostatic compensating valve, this technology senses the temperature of the water to adjust the mix of hot and cold water. This maintains a safe, comfortable water temperature whether the fluctuation is due to a change in the pressure or the temperature of the incoming hot and cold water supplies.

■ **Thermal shock.** A large and rapid change in the water temperature. Thermal shock is a particular concern for showers where rapid changes in the temperature of the water can lead to scalding, as well as increased risk of injuries due to slips and falls. Technologies to prevent thermal shock include pressure-balance and thermostatic shower valves.

■ **Universal design.** Universal design should be accessibility that is not apparent and, at the same time, can accommodate a wide variety of people of all ages and statures. It allows access to a richer life by eliminating disability by design. This thoughtful approach to space and borders allows the maximum number of people to use the widest variety of products in their homes for the greatest length of time.

# Review Questions

1. What do GPM and GPF mean?

   _____

2. What is a lavatory?

   _____

3. What is vitreous china and where is it used?

   _____

4. What does leach mean?

   _____

5. What is the Latin word for lead?

   _____

6. What does Proposition 65 deal with?

   _____

7. What is meant by barrier-free?

   _____

8. What is thermal shock?

   _____

# 3 | Plumbing Codes and Standards

## Performance Objectives

After studying this chapter, you will:

- Be familiar with the various codes encountered in the plumbing trade.

- Understand why these various Standards and Codes are necessary.

- Be able to describe which code applies to which type of pipe.

- Know how soil pipe is cut and installed.

- Be able to answer the review questions at the end of the chapter.

# Model Codes

## NSPC

National Association of Plumbing Heating Cooling Contractors, Falls Church, VA

## The National Standard Plumbing Code

The National Standard Plumbing Code is used in Maryland, New Jersey, and some cities.

## IAPMO

International Association of Plumbing and Mechanical Officials, Walnut, CA

## Uniform Plumbing Code

The Uniform Plumbing Code is used in Western United States.

## ICC

International Code Council, Whittier, CA

## International Plumbing Code

The IPC is a new plumbing code developed by the BOCA, ICBO, and SBCCI membership. The latest printing is 2017.

# Specifications

## ASTM Specifications

### ASTM A 674-05

Standard practice for polyethylene encasement for ductile iron pipe for water or other liquids. This specification is updated and revised every five years. The number following A 674 indicates the year of last revision. This specification and ANSI A21.5 are referenced for external corrosion protection for cast iron installed underground. The ASTM A 74-98 and A 888-98 contain information contained in this standard.

### ASTM A 74-06

Standard specification for cast iron soil pipe and fittings. This specification is reviewed and revised every five years. The number following A 74 indicates the year of latest revision.

### ASTM A 888-06

Standard specification for hub-less cast iron soil pipe and fittings for sanitary and storm drain, waste, and vent piping applications. This specification was first printed in 1991 and will be reviewed and revised every five years. The number following A 888 indicates year of latest revision.

### ASTM C 1277-06

Standard specification for shielded couplings joining hub-less cast iron soil pipe and fittings. New standard for shielded couplings used to join hub-less cast iron pipe and fittings. CISPI 310 couplings meet this standard as do most heavy duty couplings.

### ASTM C 564-03a

Standard specification for rubber gaskets for joining cast iron soil pipe and fittings. This specification is reviewed and revised every five years. The number following C 564 indicates the year of last revision.

## CISPI Specifications

### CISPI 301-05

Standard specification for hub-less cast iron soil pipe and fittings for sanitary and storm drain, waste, and vent piping applications. This specification is reviewed and revised every five years. The number following 301 indicates the last year of revision.

### CISPI 310-04

Specification for coupling for use in connection with hub-less cast iron soil pipe and fittings for sanitary and storm drain, waste, and vent piping applications. This specification is reviewed and revised every five years. The number following 310 indicates the last year of revision.

## CISPI Designation HS 74-86

Specification for cast iron soil pipe and fittings for hub and spigot sanitary and storm drain, waste, and vent piping applications. This seldom-referenced specification modified ASTM A-74 by making pressure testing optional and appropriate collective marks optional.

## CISPI Designation HSN-85

Specification for neoprene rubber gaskets for hub and spigot cast iron soil pipe and fittings. This seldom-referenced specification covers preformed rubber gaskets used to seal joints in hub and spigot cast iron soil pipe and fittings. The most common specification referenced today is ASTM C 564.

## Other Specifications

### CS 188-66

Specification for cast iron soil pipe and fittings. Issued by the U.S. Department of Commerce in 1966, it includes extra heavy and service pipe. This specification was another version of ASTM A 74 and is occasionally referenced in some city plumbing codes. It is no longer a current specification.

### CSA B 602

Mechanical couplings for drain, waste, and vent pipe and sewer pipe. This specification, issued by the Canadian Standards Association, covers couplings used in transition between dissimilar materials and the same materials. CSA B 602 is not referenced as a coupling standard by any model plumbing code, except for use as underground transition couplings. The standard is in metric units only.

### CSA B 70

Cast iron soil pipe, fittings and means of joining. This specification, issued by the Canadian

Standards Association, covers cast iron soil pipe and fittings. Although the title refers to joining, the reader is referred to CSA B 602. This standard is in metric units only.

### GEGS-15410

Corps of Engineers Guide Specification, Military Construction. This specification includes references to CISPI 301 and ASTM A 74 for soil pipe and fittings, and CISPI 310 for hub-less couplings. This specification is also used by the General Services Administration.

### WWP 401F

Federal specification for pipe and pipe fittings, cast iron soil. This specification, issued by U.S. Department of Defense, refers back to CISPI 301 and ASTM A 74 for the material specifications. This specification includes information about packaging, compliance reports, etc. It is probable that this standard will be deleted by federal agencies. CISPI 301, CISPI 310, and ASTM A 74 are currently referenced by the Department of Defense and the Army Corp of Engineers.

## American Society for Testing and Materials (ASTM) Standard Specifications

### ASTM A 74

Standard specifications for Hub and Spigot Cast Iron Soil Pipe and Fittings

### ASTM A 888

Standard specifications for Hub-less Cast Iron Soil Pipe and Fittings

### ASTM C 564

Standard specifications for Rubber Gaskets for Cast Iron Soil Pipe and Fittings

## Cast Iron Soil Pipe Institute (CISPI) Standard Specifications

### CISPI 301

Hub-less Cast Iron Soil Pipe and Fittings for Sanitary and Storm Drain, Waste and Vent Piping Applications (see Figures 3-1 and 3-2).

### CISPI 310

Couplings for use in connection with Hub-less Cast Iron Soil Pipe and Fittings for Sanitary and Storm Drain, Waste, and Vent Piping applications.

*(Information courtesy of the Cast Iron Soil Pipe Institute.)*

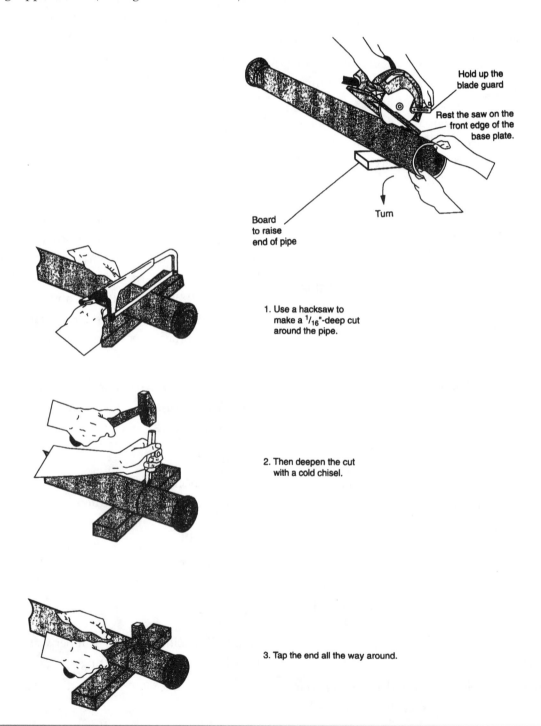

Hold up the blade guard

Rest the saw on the front edge of the base plate.

Board to raise end of pipe

Turn

1. Use a hacksaw to make a $^1/_{16}$"-deep cut around the pipe.

2. Then deepen the cut with a cold chisel.

3. Tap the end all the way around.

**Figure 3-1** Using different methods to cut cast iron pipe.

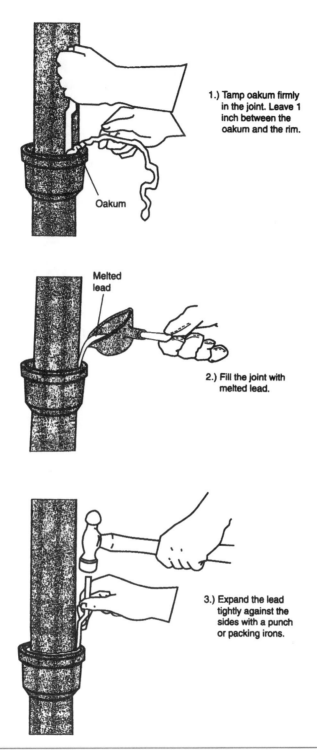

1.) Tamp oakum firmly in the joint. Leave 1 inch between the oakum and the rim.

Oakum

Melted lead

2.) Fill the joint with melted lead.

3.) Expand the lead tightly against the sides with a punch or packing irons.

**Figure 3-2** Steps in preparing a cast iron joint.

## Review Questions

1. Where is the National Standard Plumbing Code located?

   _____

2. Where is the Uniform Plumbing Code used?

   _____

3. What does ASTM specify?

   _____

4. Identify the following abbreviations:
   a   ASTM
   b.  CISPI
   c.  IAPMO
   d.  IPC

# 4 | Bathtubs

## Performance Objectives

After studying this chapter, you will:

- Understand why there are so many types of bathtubs.

- Be able to identify the various types of bath tubs.

- Know how to install a bathtub in a new house.

- Be able to describe how bathtubs are important in a bathroom.

- Be able to explain why a corner bathtub is ideal for a given location.

- Be able to define the bathtub by name.

- Be able to answer the review questions at the end of the chapter.

Bathtubs are available as single units and as combinations (shower/tub). There are also some specially equipped bathtubs. Roughing-in dimensions are given in Figures 4-1 (A, B). Note that the waste location is 1¼ inches (38 mm) off the rough wall on the centerline of the waste. The P-trap top should be no higher than 7 inches (18 cm) below floor level. Put the drain piece on last.

The shower rod location is 76 inches (193 cm) high and 27 inches (69 cm) off the finished wall or against the outside tub rim. Figure 4-2 gives the exploded view of the typical tub trip waste overflow.

Figure 4-3 provides instructions on installing the single handle tub and shower set.

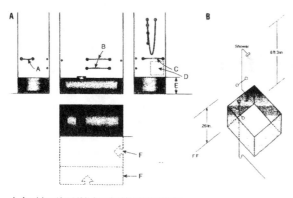

A  A grab bar at least 12 in. long should be installed on the head end wall at the front edge of the bathtub.

   Diameter or width of grab bars should be 1¼ in. to 1½ in. with a 1½ -in. space between the grab bar and the wall.

B  A grab bar 33 in. to 26 in. above the floor and at least 24 in. long should be installed on the back wall, 24 in. maximum from the head end wall and 12 in. maximum from the foot end wall. Another grab bar of equal length should be installed 9 in. above the rim of the bathtub.

C  A grab bar at least 24 in. long should be installed on the foot end wall at the front edge of the bathtub.

D  Control area.

E  Rim of the bathtub should be 17 in. to 19 in. above the floor.

F  30 in. × 60 in. minimum clear floor space for a parallel approach to a bathtub and 48 in. and times; 60 in. minimum for a forward approach.

**Figure 4-1**  (A) Architect's roughing-in dimensions for bathtub placement. (B) Tub view, roughing-in.

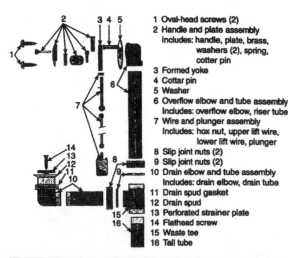

1  Oval-head screws (2)
2  Handle and plate assembly
   Includes: handle, plate, brass, washers (2), spring, cotter pin
3  Formed yoke
4  Cottar pin
5  Washer
6  Overflow elbow and tube assembly
   Includes: overflow elbow, riser tube
7  Wire and plunger assembly
   Includes: hex nut, upper lift wire, lower lift wire, plunger
8  Slip joint nuts (2)
9  Slip joint nuts (2)
10 Drain elbow and tube assembly
   Includes: drain elbow, drain tube
11 Drain spud gasket
12 Drain spud
13 Perforated strainer plate
14 Flathead screw
15 Waste tee
16 Tall tube

**Figure 4-2**  Exploded view. Typical tub trip waste overflow.

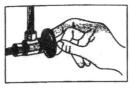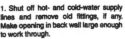

1. Shut off hot- and cold-water supply lines and remove old fittings, if any. Make opening in back wall large enough to work through.

2. Connect water supply lines—hot water to left and cold water to right.
NOTE: Use Teflon tape or pipe compound on all threaded connections. When sweating a valve, all plastic and rubber components should be removed from the casting to protect them from the heat.

3. Install ½-in. shower supply line (not included) to center connection on valve assembly and to a ½-in. pipe elbow (not included) at desired shower height. For shower installation only, plug bottom outlet.

4. Assemble ½-in. tub spout supply line to ½-in. elbow (not included). Screw this assembly into bottom opening of valve assembly. Pipe elbow should reach wall opening just above tub. For tub installation only, plug top outlet.

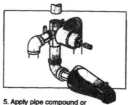

5. Apply pipe compound or Teflon tape to tub spout nipple (not included). Attach tub spout to elbow. Tighten by hand.

6. Place flange over long end of shower arm and apply pipe compound or Teflon tape to threads on end. Screw into elbow at the shower supply line. The shower head can now be attached to shower arm.
CAUTION: Use cloth over shower head nut to prevent scratching during installation.

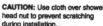

7. Slip escutcheon over valve cover and screw into wall. Slip acrylic handle onto stem and screw securely into place using the screws provided. Snap on index button. Adjust the temperature limit stop; see CAUTION statement in step 6.

8. To flush water lines, let faucet run fully open through tub spout (unless shower only) for 1 minute each in the hot and cold positions.

**Figure 4-3**  Installation instructions for the single-handle tub and shower set. (American Standard Brands)

Specifications for American Standard tubs are given in Table 4-1 through Table 4-3:

- Rectangular Tub Specifications (Table 4-1)
- Oval Tub Specifications (Table 4-2)
- Corner Tub Specifications (Table 4-3)

The whirlpool tub is a little more complicated than the regular fiberglass tub nailed into position after leveling.

- Figure 4-4A shows views of the rectangular whirlpool tub.

**Table 4-1**   Specifications of the Rectangular Bathtub

| Model | Dimensions (Length x Width x Height) | Drain/ Overflow Dimensions | Cutout | Total Weight/ Floor Loading | Operating Gallonage | Produce Weight | Skirt and Mounting |
|---|---|---|---|---|---|---|---|
| Cypress 5 ft x 32 in. | 60 in. x 32 in. x 19¼ in. | 15¾ in./9¼ in. | 58 in. x 30 in. | 604 lb/45 lb/ft² | 43 U.S. gal. (162.77 L) | 79 lb | Optional |
| Savoy 5 ft x 32 in. | 60 in. x 42 in. x 18¼ in. | 14³⁄₁₆ in./8¾ in. | 58 in. x 40 in. | 789 lb/45 lb/ft² | 46 U.S. gal. (174 13 L) | 97 lb | Optional |
| Patriot 6 ft x 36 in. | 72 in. x 36 in. x 19¼ in. | 15¾ in./9½ in. | 70 in. x 34 in. | 745 lb/42 lb/ft² | 52 U.S. gal. (196.84 L) | 97 lb | Optional |
| Emblem 6 ft x 42 in. | 72 in. x 42 in. x 20½ in. | 15¾ in./8½ in. | 70 in. x 34 in. | 885 lb/42 lb/ft² | 63 U.S. gal. (238.48 L) | 99 lb | Optional |
| Berkeley 5 ft x 36 in. | 60 in. x 36 in. x 19¼ in. | 15¾ in./9¼ in. | 58 in. x 34 in. | 643 lb/37 lb/ft² | 49 U.S. gal. (185.49 L) | 85 lb | Optional |
| Dakota 6 ft x 48 in. | 72 in. x 48 in. x 18¼ in. | 14½ in./11⅜ in. | 70 in. x 46 in. | 885 lb/37 lb/ft² | 74 U.S. gal. (280.12 L) | 118 lb | Optional |

For all units:
Motor/pump: 115 VAC, 3450 rpm/7.8 amps, 60 Hz, single phase.
Electrical requirements: 115 VAC, 15 amps, 60 Hz. Requires a dedicated circuit.

*Reproduced with permission of American Standard Brands.*

**Table 4-2**   Specifications of the Oval Bathtub

| Model | Dimensions (Length x Width x Height) | Drain/Overflow Dimensions | Cutout | Total Weight/ Floor Loading | Operating Gallonage | Product Weight | Skirt and Mounting |
|---|---|---|---|---|---|---|---|
| Laguna 5 5-ft oval | 62 in. x 43 in. x 18¾ in. | 15⅛ in./9½ in. | Template provided | 788 lb/58 lb/ft² | 50 U.S. gal. (189.27 L) | 96 1b | Not available |
| Laguna 6 6-ft oval | 72 in. x 42 in. x 20½ in. | 16 in./11⅜ in. | Template provided | 880 lb/52 lb/ft² | 68 U.S. gal. (257.41 L) | 105 lb | Not available |

For all units:
Motor/pump: 115 VAC, 3450 rpm/7.8 amps, 60 Hz, single phase.
Electrical requirements: 115 VAC, 15 amps, 60 Hz. Requires a dedicated circuit.

*Reproduced with permission of American Standard Brands.*

**Table 4-3**  Specifications of the Corner Bathtub

| Model | Dimensions (Length x Width x Height) | Drain/Overflow Dimensions | Cutout | Total Weight/ Floor Loading | Operating Gallonage | Product Weight | Skirt and Mounting |
|---|---|---|---|---|---|---|---|
| Triangle 5-ft corner, 5-ft DE, Esquina | 60 in. x 60 in. x 19¾ in. | 14⅞ in./11 in. | Template provided | 770 lb/41 lb/ft² | 47 U.S. gal. (177.91 L) | 119 1b | Optional |
| Triangle II 60 in x 60 in. | 60 in. x 60 in. x 22 in. | 17¾ in./10⅞ in. | Template provided | 872 lb/40 lb/ft² | 62 U.S. gal. (234.70 L) | 122 lb | Not available |

For all units:
Motor/pump: 115 VAC, 3450 rpm/7.8 amps, 60 Hz, single phase.
Electrical requirements: 115 VAC, 15 amps, 60 Hz. Requires a dedicated circuit.

*Reproduced with permission of American Standard Brands.*

- Figure 4-4B shows the side, end and drain overflow views of the rimless oval bathtub.
- Figure 4-4C shows the views of the corner bathtub.

Roughing-in references for whirlpool tubs are shown in Figure 4-5. Figure 4-6 illustrates U-frame skirt mounting detail.

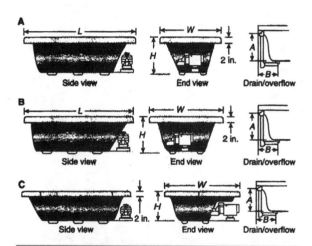

**Figure 4-4**  (A) Rectangular whirlpool tub, side, end and drain/overflow views. (B) Views of the rimless oval bathtub. (C) Views of the corner bathtub. (American Standard Brands)

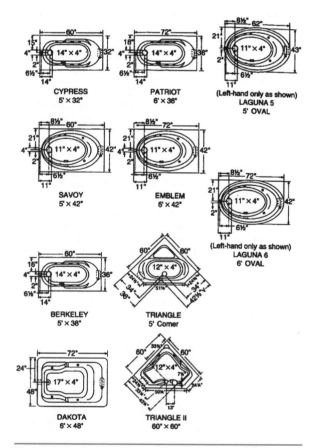

**Figure 4-5**  Roughing-in references for whirlpool tubs. (American Standard Brands)

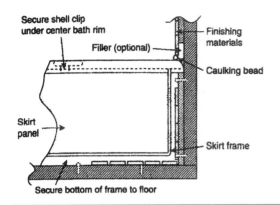

**Figure 4-6**   Mounting details for a U-frame skirt. (American Standard Brands)

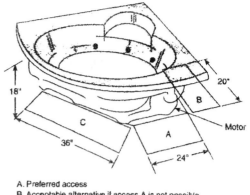

A. Preferred access
B. Acceptable alternative if access A is not possible
C. Optional access for accessory equipment

**Figure 4-8**   Service access 5-foot corner triangle tub. (American Standard Brands)

## Service Access

If you have a partially or fully sunken installation, space should be allowed for access later if something goes wrong. Figure 4-7 shows the service access without the skirt. It is the responsibility of the installer to provide sufficient service access. The recommended minimum dimensions allowable for service to the tub are given in Figures 4-7 and 4-8.

## Electrical Connections

Note the location of the electrical outlet in Figure 4-9. Access should be provided to this area. The corner bath installation of the electrical outlet is shown in Figure 4-10.

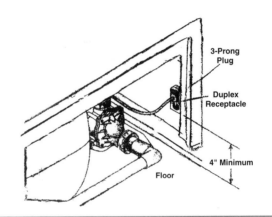

**Figure 4-9**   Electrical connection for side/end drain bathtubs. (American Standard Brands)

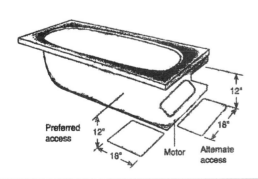

**Figure 4-7**   Service access without the skirt. (American Standard Brands)

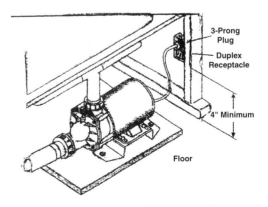

**Figure 4-10**   Corner bathtub electrical connection. (American Standard Brands)

## Cast Iron Tubs

Porcelain-enameled cast iron tubs and their dimensions are shown in Figure 4-11. These are a little heavier than the plastic tubs and will probably utilize the service of two installers to make sure the porcelain enamel is not chipped.

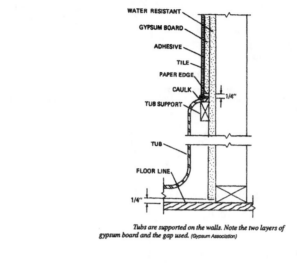

*Tubs are supported on the walls. Note the two layers of gypsum board and the gap used. (Gypsum Association)*

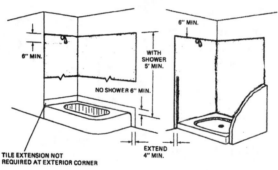

**Figure 4-11**   A recessed porcelain-enameled cast-iron tub with roughing-in dimensions. (Crane)

# Review Questions

1. What does it mean to rough-in a bathtub?

   _____

2. How many types of bathtubs are there?

   _____

3. How do you identify the various types of tubs?

   _____

4. How do you estimate how large a bathtub is needed for a given location?

   _____

5. What is the difference between a cast iron tub and one of another material?

   _____

# 5 | Drains, Vents, and Vent Piping

## Performance Objectives

After studying this chapter, you will:

- Understand how a sanitary drain is connected to a street sanitary sewer.

- Know the proper size pipe to use for drains and vent.

- Know how a drainage system works.

- Be able to explain how vents are selected and produced.

- Know which pipe fittings should be used and how to make a vent system.

- Be able to answer the review questions at the end of the chapter.

Each fixture in the water supply system terminates at a plumbing fixture. After the supply water has been drawn and used, it utilizes the sanitary drainage system for disposal. The primary purpose of the drainage system is to dispose of fluid waste and organic matter as quickly as possible (see Figure 5-1).

The sanitary drainage system relies on gravity for its operation (see Figure 5-2). Its pipes are much larger than the water supply lines that

operate under pressure. The drainage lines are sized according to their location in the system and the total number of types of fixtures served. Check the plumbing code for allowable materials, such as pipe, pipe size, restrictions on the length and slope of horizontal runs, and the types and number of turns allowed in the piping. Table 5-1 provides typical fixture allowances in relationship to drainage pipe sizes.

**Table 5-1**   Typical Fixture Allowances—Drainage Pipe Sizes

| Fixture Type | Drainage Fixture Unit Value as Load Factors | Minimum Size of Trap (Inches) |
|---|---|---|
| Automatic clothes washers, commercial | 3 | 2 |
| Automatic clothes washers, residential | 2 | 2 |
| Bathroom group as defined in Section 202 ( 1.6 gpf water closet) | 5 | — |
| Bathroom group as defined in Section 202 (water closet flushing greater than 1.6 gpf) | 6 | — |
| Bathtub (with or without overhead shower or whirlpool attachments) | 2 | 1½ |
| Bidet | 1 | 1¼ |
| Combination sink and tray | 2 | 1½ |
| Dental lavatory | 1 | 1¼ |
| Dental unit or cuspidor | 1 | 1¼ |
| Dishwashing machine, domestic | 2 | 1½ |
| Drinking fountain | ½ | 1¼ |
| Emergency floor drain | 0 | 2 |
| Floor drains | 2 | 2 |
| Kitchen sink, domestic | 2 | 1½ |
| Kitchen sink, domestic with food waste grinder and/or dishwasher | 2 | 1½ |
| Laundry tray (1 or 2 compartments) | 2 | 1½ |
| Lavatory | 1 | 1¼ |
| Shower | 2 | 1½ |
| Sink | 2 | 1½ |
| Urinal | 4 | |
| Urinal, 1 gallon per flush or less | 2 | |
| Wash sink (circular or multiple) each set of faucets | 2 | 1½ |
| Water closet, flushometer tank, public or private | 4 | |
| Water closet, private (1.6 gpf) | 3 | |
| Water closet, private (flushing greater than 1.6 gpf) | 4 | |
| Water closet, public (1.6 gpf) | 4 | |
| Water closet, public (flushing greater than 1.6 gpf) | 6 | |

For S1:  1 inch = 25.4 mm, 1 gallon = 3.785 L.

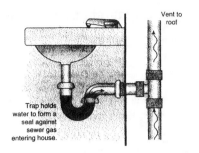

**Figure 5-1** How a trap works. (Genova Products Inc.)

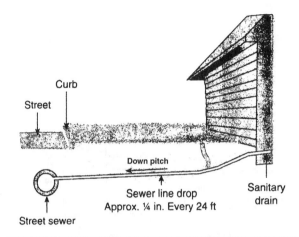

**Figure 5-2** Sanitary drain connected to a street sanitary sewer. (Adapted from City of Reno, Nevada Public Works Maintenance & Operations Sewer & Storm Drain, www.reno.gov/index.aspx?page=559)

## Drainage Lines

Cast iron or plastic is used for drainage lines. Cast iron was the traditional material for drainage piping for decades. It may have been hub-less or bell and spigot joints and fitting. Two types of plastic pipe suitable for drainage lines are polyvinyl chloride (PVC) and acrylonitrile-butadiene styrene (ABS). Some codes permit the use of galvanized wrought iron or steel.

## Drains

Any waste disposal system requires drains, and plumbers are drain specialists.

They are concerned with the disposal of sewage, waste, and run-off from the roofs. See Table 5-2 for typical conductor pipe (down spout) sizes. Roof area determines the size of the

**Table 5-2** Typical Drain Pipe Sizes vs. Vertical Roof Area

| PIPE DIMENSIONS | | | | |
|---|---|---|---|---|
| Dimensions of Schedule 40 (Standard Weight) Steel and Wrought Iron Pipe | | | | |
| Nominal Diameter, Inches | Actual Inside Diameter, Inches | Actual Outside Diameter, Inches | Circumference, Outside, Inches | Weight per Foot, Pounds |
| ⅛ | 0.269 | 0.405 | 1.27 | 0.25 |
| ¼ | 0.364 | 0.540 | 1.69 | 0.43 |
| ⅜ | 0.493 | 0.675 | 2.12 | 0.57 |
| ½ | 0.622 | 0.840 | 2.65 | 0.86 |
| ¾ | 0.824 | 1.050 | 3.29 | 1.14 |
| 1 | 1.049 | 1.315 | 4.13 | 1.68 |
| 1¼ | 1.380 | 1.660 | 5.21 | 2.28 |
| 1½ | 1.610 | 1.900 | 5.96 | 2.72 |
| 2 | 2.067 | 2.375 | 7.46 | 3.66 |
| 2½ | 2.469 | 2.875 | 9.03 | 5.80 |
| 3 | 3.068 | 3.500 | 10.96 | 7.58 |
| 3½ | 3.548 | 4.000 | 12.56 | 9.11 |
| 4 | 4.026 | 4.500 | 14.13 | 10.80 |
| 5 | 5.047 | 5.563 | 17.47 | 14.70 |
| 6 | 6.065 | 6.625 | 20.81 | 19.00 |
| 8 | 7.981 | 8.625 | 27.09 | 28.60 |
| 10 | 10.020 | 10.750 | 33.77 | 40.50 |
| 12 | 11.938 | 12.750 | 40.05 | 53.60 |
| 14 | 13.126 | 14.000 | 47.12 | 63.30 |
| 16 | 15.000 | 16.000 | 53.41 | 82.80 |
| 18 | 16.876 | 18.000 | 56.55 | 105.00 |

conductor pipe in a building's drainage system (see Figures 5-3 to 5-9).

## Pipe

Pipe is available in straight length sections from 12 feet to 20 feet (3.66m to 6.10m) long. Standard wrought iron or steel piping up to 12 inches in diameter is classified by its nominal inside diameter. Actual inside diameter may vary for a given nominal size (such as standard pipe, heavy pipe, or extra heavy pipe); classification depends on pipe weight factors. External diameter is normally the same for all three weights. Pipe above 12 inches (30.48 cm) is classified by its actual diameter.

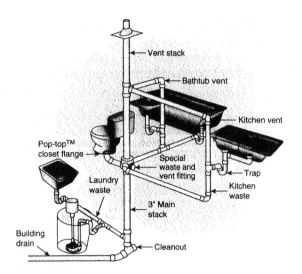

**Figure 5-3**  Drain, waste, and vent systems. (Genova Products, Inc.)

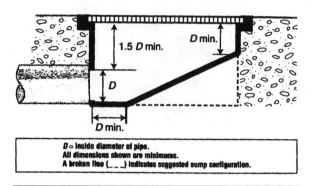

**Figure 5-6**  Main drain for a swimming pool. (North Carolina Department of Environmental and Natural Resources, www.deh.enr.state.nc.us/eha/images/pti/figure2.png)

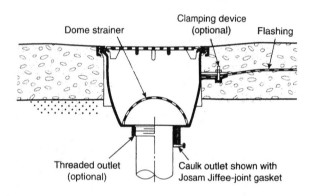

**Figure 5-4**  Diagram of a typical floor drain installation.

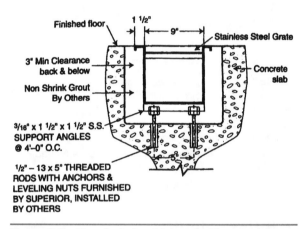

**Figure 5-7**  Trench drain with interlocking vandal proof grating used in swimming pools. (Superior Swim Systems, www.superiorswimsystems.com/gutters.php)

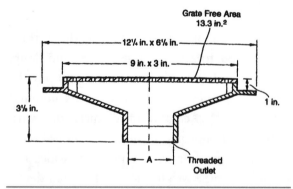

**Figure 5-5**  Typical swimming pool scum-gutter drain. (Jay R. Smith Mfg. Co.)

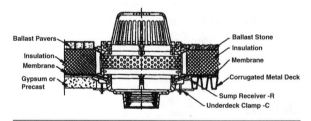

**Figure 5-8**  Conductor pipe drain for a typical flat roof. (Jay R. Smith Mfg. Co.)

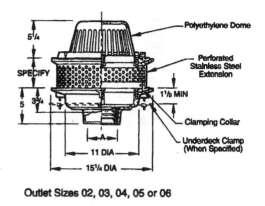

**Outlet Sizes 02, 03, 04, 05 or 06**

**Figure 5-9** Installation diagram for a flat-roof conductor-pipe. (Jay R. Smith Mfg. Co.)

Brass and copper piping are classified by the same nominal sizes as iron pipe with two weights for each size—extra strong and regular.

## Pipe Fittings

Pipe fittings provide continuity by using couplings, nipples, and reducers. The use of tees, crosses, and elbows is primarily to change direction of a run of piping. Caps are used to close open pipes, plugs are used to close open fittings, and bushings are used to reduce the size of an opening.

Unions are available to make connections convenient and easily unmade. Screwed unions are of three-piece design. Two pieces are screwed to the ends of the pipes being connected. The threads draw them together by screwing onto the first piece and bearing against the shoulder of the second. Flow in a piping system is regulated by valves that are specified by type, material, size, and working pressures.

## Water Distribution Systems

The water distribution system, located inside a building, supplies water to the plumbing and its fixtures. Water requirements to each fixture are supplied from the water main with reduced-diameter pipe connections. Two types of water

supply systems are in general use: the down-feed and the up-feed.

- In a down-feed system, the supply tank is located on the roof or high point on the inside of a building. Gravity causes the water to flow to the fixtures after it is mechanically pumped to the supply tank.
- In an up-feed system, water is supplied from an elevated storage tank or water tower. The lowest elevation point of the water tank is located so that it is above the highest point of the water distribution system. Water fills the supply system lines as it seeks its water normal or natural level.

Cold and hot water supply systems usually rely on copper, brass, wrought iron or galvanized steel pipes. Fittings are made of the same materials as the pipe. The type of material to be used in a building is usually specified on the prints or separate sheets of specifications.

## Vents (Figures 5-10 to 5-19)

The key points regarding vents are:

- **Waste water pipe.** Waste water piping in a plumbing system carries all the other wastes except those from the toilet. Waste vents can extend through the roof. Or, they can go into the soil vent just above the highest fixture.
- Vents are used to introduce air into a plumbing system and break the suction resulting from the falling or draining waste-pipe water.
- **Vents and re-vents.** Main vents service the major or master bathroom in the house or building. Re-vents are secondary vents. They are used to vent the fixtures located more than 8 feet from the main vent. Re-vents relieve air pressures on fixtures that are vertically downstream from the water closet or toilet.

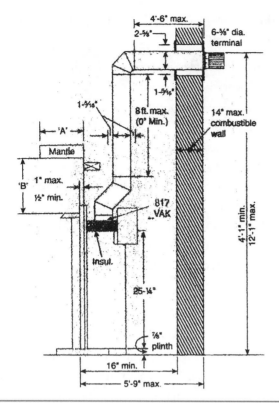

Figure 5-10   Method for vent stack offset. (Victorian Fireplace Shop)

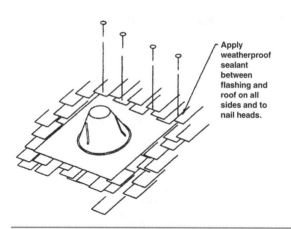

Figure 5-11   Commonly used vent increaser with roof flashing. (Simpson Dura-Vent Co., Inc.)

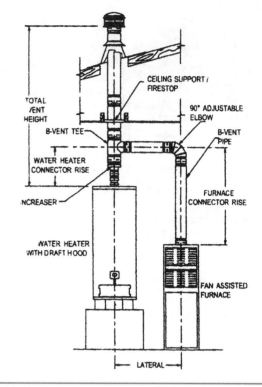

Figure 5-12   Vent increaser roof extension. (Simpson Dura-Vent Co., Inc.)

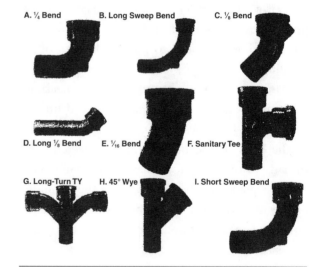

Figure 5-13   Commonly used drainage fittings. (Charlotte Pipe and Foundry Company)

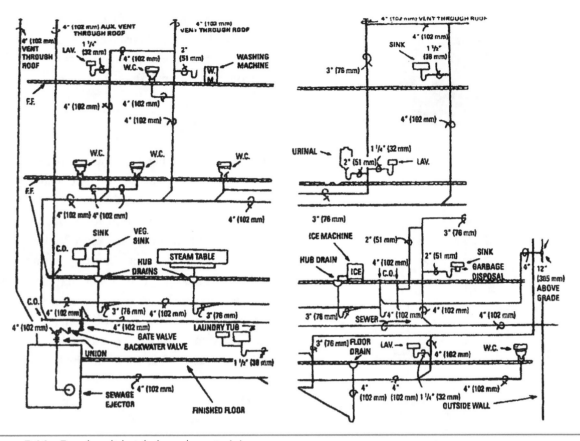

**Figure 5-14**   Free-hand sketch: branch-vent piping.

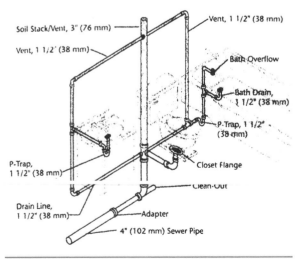

**Figure 5-15**   Branch-vent installation. (National Kitchen & Bath Association, www.NKBA.org)

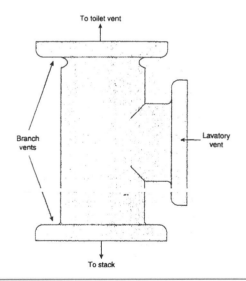

**Figure 5-16**   Tapering reducer tee. (Anvil International)

**Figure 5-17**    Cast iron vandal-proof hooded vent. (Josam Mfg. Co.)

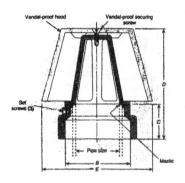

**How a vent system operates, in three steps. (A) Without a vent, trap water is siphoned off. Too little water remains in the trap to block sewer gases. (B) Sewer gases enter the interior of the house. (C) A vent allows air to rush in and create a gas seal by preventing trap water from siphoning off.**

**Figure 5-18**    Installation diagram: cast iron-vandal-proof hooded vent cap. (Josam Mfg. Co.)

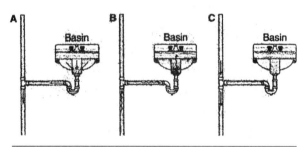

**Figure 5-19**    ABC steps: How a vent system operates. (Carson, Dunlop, and Associates, Ltd.)

- **Soil vents and soil pipes.** Soil pipe is defined as that part of a system that carries away the solid and liquid wastes from the water closet (toilet) or similar fixtures. The soil vent is the vent pipe that extends through the roof. It relieves the unequal air pressure that results from waste falling or running

down the pipe. Also, a vent prevents traps from siphoning themselves dry. By introducing air, they break the suction that results from the falling water.

## Types of Vents

Vents can be further classified into two types: wet vents and dry vents. Wet vents usually have two purposes. A wet vent can serve as a drain for one fixture and as a vent for another. A good example is when toilets are vented with a lavatory. This means you place a fitting within a prescribed distance from the toilet that also serves as a drain for the lavatory. From the drain it goes to the lavatory. The wet vent becomes a dry vent once it extends above the trap arm. Distances and specifications are found in the local plumbing codes.

Dry vents are used for many fixtures. These are vents that do not receive the drainage discharge of a fixture. The pipes only carry air and are called dry vents.

Some of the many types of dry vents include:

- Circuit vents
- Common vents
- Individual vents
- Relief vents
- Vent stacks

## Vent Piping

In home plumbing, vent piping consists of two separate systems: the water supply system and the water disposal system. Water in the water supply system is under about 50 psi (345 kPa). Water supply pipes are often smaller in diameter, but still carry drain-waste-vent (DWV) system flows by gravity. That means the piping and fittings in the DWV system are larger in diameter. Larger pipes are needed to carry the required flow without clogging or backing up. Both systems will operate safely if properly designed. For years, cast iron was the material of choice for these two types of systems not only in houses, but also in industry and business.

Modern homes either use PVC (plastic) pipe or copper pipe instead of cast iron. Some newer types of plastic covered piping recently replaced copper in these applications. Cast iron pipes have been around for decades. They are usually found in older homes, as well as commercial and industrial installations. Cast iron piping requires a practitioner with a great deal of experience in working with lead, Table 5-3, while Table 5-4 shows some commonly used vent sizes and lengths. Pipe calculations are of total transverse area for the vent pipes.

**Table 5-3**   Calculations of Total Transverse Area for Vent Pipe

| Internal Transverse Area (in.²) | Vent Pipe Size (in.) |
|---|---|
| 2.036 | 1½ (lavatory) |
| 3.355 | 2 (water closet) |
| 5.391 (total) | — |

Data from the Engineering Toolbox, Outside Diameter, Identification, Wall Thickness, Inside Diameter (www.engineering toolbox.com/steel-pipes-dimensions-d_43.html)

**Table 5-4**   Some Commonly Used Vent Sizes and Length

| Soil or Waste Stack Diameter (in.) | Number of Vented Fixture Units | Vent Diameter (in.) | | | | | |
|---|---|---|---|---|---|---|---|
| | | 1¼ | 1½ | 2 | 2½ | 3 | 4 |
| | | Maximum Developed Length (ft.) | | | | | |
| 1¼ | 2 | 30 | — | — | — | — | — |
| 1½ | 8 | 50 | 150 | — | — | — | — |
| | 10 | 30 | 100 | — | — | — | — |
| 2 | 12 | 30 | 75 | 200 | — | — | — |
| | 20 | 26 | 50 | 150 | — | — | — |
| 2½ | 42 | — | 30 | 100 | 300 | — | — |
| 3 | 10 | — | 42 | 150 | 360 | 1040 | — |
| | 21 | — | 32 | 110 | 270 | 810 | — |
| | 53 | — | 27 | 94 | 230 | 680 | — |
| | 102 | — | 25 | 86 | 210 | 620 | — |
| 4 | 43 | — | — | 35 | 85 | 250 | 980 |
| | 140 | — | — | 27 | 65 | 200 | 750 |
| | 320 | — | — | 23 | 55 | 170 | 640 |
| | 540 | — | — | 21 | 50 | 150 | 580 |

Data from Ohio Administrative Code, 4101:3 Board of Building Standards: Ohio Plumbing Code, Chapter 4101:3-9-01 Vents. Table 916.1: Size and Developed Length of Stack Vents and Vent Stacks [http://codes.ohio.gov/oac/4101:3-9-01]. Accessed May 20 2010.

## Review Questions

1. What are the two types of vent systems?

   _____

2. How is a vent system constructed?

   _____

3. How do you know where to place vents?

   _____

4. How do vents enter outside space in a house?

   _____

5. What type of material is the pipe made of in the present-day home?

   _____

6. What does DWV stand for in plumbing?

   _____

7. What is PVC pipe made of?

   _____

# 6 | Fittings

## Performance Objectives

After studying this chapter, you will:

- Be able to identify the various types of piping used in water systems.

- Know what PEX means.

- Know what the word polyethylene means.

- Understand the meaning of PE piping.

- Know what CPVC means.

- Know what HDPE pipe is and where to use it.

- Be able to identify CPVC from the newer Fernco fittings.

- Be able to identify and utilize various fittings, both plastic and metal.

- Be able to answer the review questions at the end of the chapter.

Every type of pipe or tubing has its fittings. Some joints must be soldered with hot metal solder or with plastic cold joint cement. The variety of different types of fittings makes choosing the right one difficult. However, with practice the plumber is capable of identifying each and every one by name and by sight.

There are three primary types of pipes used in above-grade water systems:

- PEX (cross-linked polyethylene)
- Copper
- CPVC (chlorinated polyvinyl chloride)

The PEX is very similar to polyethylene in appearance, but because of cross linking is a thermoset material—it does not melt. It is highly flexible and easily coiled. PEX is approved in all North American model plumbing codes for hot and cold potable water distribution systems.

CPVC, in addition to its use in domestic plumbing systems, is used extensively in fire sprinkler systems. It has a service life of over 35 years, outstanding corrosion resistance, low flame spread and low smoke emission levels that make it a good choice for this service.

The PEX-AL-PEX composite capitalizes on the corrosion and chemical resistance of plastic and pressure capacity of metal by laminating the aluminum layer between layers of plastic. The resulting tubing is non-corroding, bendable for form stability, flexible, and resists most acids, salt solutions, alkalis, fats, and oils.

## PE Piping

PE piping has been used for a variety of applications for more than 50 years. It has been accepted for its overall toughness and durability. PE pipe has a variety of installation methods that are employed to expand its use at a quickening rate.

PE piping can be used as a solid wall pipe in potable water and natural gas lines. It is also used in gravity sewers and can handle most chemicals. PEX pipe is used for plumbing and heating and corrugated PE pipe is used for drainage.

The Plastics Pipe Institute (PPI) has a handbook of Polyethylene (PE) pipe with details needed for design purposes. Visit the PPI at www.plasticpipe.org.

## Heat Fusion

One of the reasons High Density Polyethylene Pipe (HDPE) is chosen over heavier steel pipe is its ability to be joined by heat fusion. The process is one that can be taught to the inexperienced person in a short time. Heat fusion of polyethylene pipe is just a matter of melting the ends of two piping components together. The necessary equipment uses normal electrical sources, usually from a standard generator. The process involves cleaning the pipe, preparing the surface, heating with a designed apparatus, and joining and cooling under pressure. Fusion joining of PE pipe has been used for over 40 years in the distribution systems of natural gas utilities in North America, and has a remarkable safety record. It has been used as the predominant water pipe in Europe since the 1960s.

## Some Selected Fittings

Fittings are available for cast iron pipe, drainage, waste and vent type copper tubing of all sizes. Figure 6-1 shows various polyvinyl chloride (PVC) waste and vent fittings. Figure 6-2 shows copper DWV fittings while Figure 6-3 highlights the variety of cast iron drainage fittings. Figure 6-4 shows some of the fittings for plastic piping for hot and cold water use. Note the cans of solvent and primer. The cans have a brush or mop arrangement included and attached to the top of the can for easy removal, usage and then storage. Figures 6-5 and 6-6 show typical plastic fittings for plastic pipe (PVC and CPVC).

**Figure 6-1**  PVC waste and vent fittings. (Nimco, Inc.)

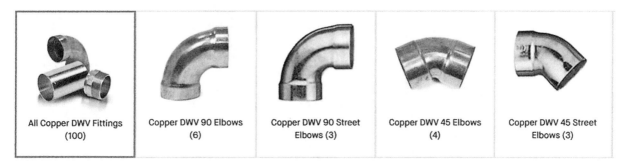

| | | | | |
|---|---|---|---|---|
| All Copper DWV Fittings (100) | Copper DWV 90 Elbows (6) | Copper DWV 90 Street Elbows (3) | Copper DWV 45 Elbows (4) | Copper DWV 45 Street Elbows (3) |

**Figure 6-2**  Copper DWV fittings. (Nimco, Inc.)

| | | | | |
|---|---|---|---|---|
| All Cast Iron Drainage Fittings (30) | Cast Iron Short Turn 90° Elbows (5) | Cast Iron Double 90° Wyes (3) | Cast Iron Drainage Tees (3) | Cast Iron Long Turn 90° Elbows (5) |

**Figure 6-3**  Cast iron drainage fittings. (Charlotte Pipe and Foundry Co.)

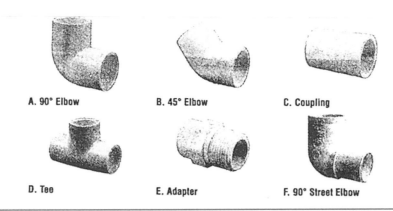

A. 90° Elbow    B. 45° Elbow    C. Coupling

D. Tee    E. Adapter    F. 90° Street Elbow

**Figure 6-4**  Hot- and cold-water pipe fittings. (Nimco, Inc.)

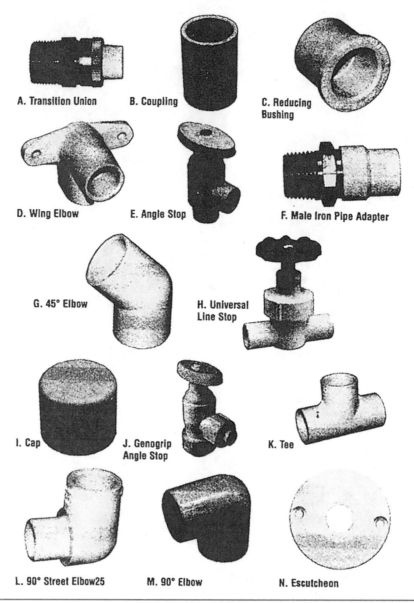

A. Transition Union　　　B. Coupling　　　C. Reducing Bushing

D. Wing Elbow　　　E. Angle Stop　　　F. Male Iron Pipe Adapter

G. 45° Elbow　　　H. Universal Line Stop

I. Cap　　　J. Genogrip Angle Stop　　　K. Tee

L. 90° Street Elbow25　　　M. 90° Elbow　　　N. Escutcheon

**Figure 6-5**　Plastic pipe water supply fittings. (Genova Products, Inc.)

# Water Distribution Pipe

A number of pipes are approved for use in distribution of water, both hot and cold, within a house or commercial building, school, and wherever people gather. The local Code requirements may vary from state to state so be sure to check which can be used in your neighborhood.

■ **Galvanized Pipe.** One of those preferred for many years. It has been mostly replaced in modern buildings with plastic and copper piping. Rarely used.

■ **Brass Pipe.** This is approved for water distribution, but is expensive and rarely used.

■ **Copper.** Most commonly used pipe for home water distribution use and elsewhere. Most commonly used types are L and M. Type K is also fine for water distribution, but it is more expensive and usually the extra thick pipe is not needed. Type M is the thinnest copper pipe and is not approved in all areas for water distribution. Type L now seems to be the standard for use with potable water distribution systems.

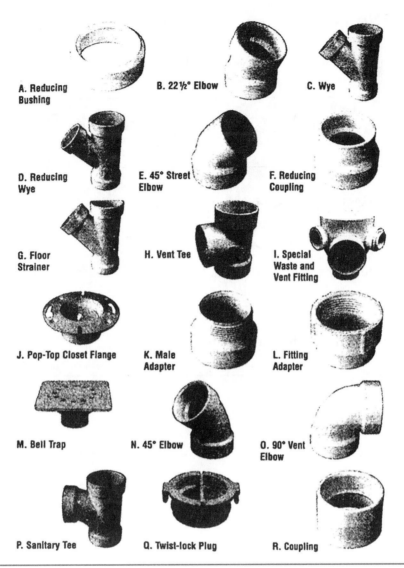

A. Reducing Bushing
B. 22½° Elbow
C. Wye
D. Reducing Wye
E. 45° Street Elbow
F. Reducing Coupling
G. Floor Strainer
H. Vent Tee
I. Special Waste and Vent Fitting
J. Pop-Top Closet Flange
K. Male Adapter
L. Fitting Adapter
M. Bell Trap
N. 45° Elbow
O. 90° Vent Elbow
P. Sanitary Tee
Q. Twist-lock Plug
R. Coupling

**Figure 6-6** Plastic pipe DWV fittings. (Genova Products, Inc.)

■ **PEX.** Easily installed and can be used in hot or cold water systems.
■ **CPVC.** Chlorinated polyvinyl chloride has been commonly used for the past 40 years. It does have some limitations so check the local code. Fittings for CPVC and PVC pipe are shown in Figures 6-5 and 6-6.

## Flexible Plastic Fittings for Drain, Waste, Venting

Fernco® flexible couplings are used for all types of in-house and sewer connections: drain, waste, vent piping, house to main, repairs, cut-ins conductor, roof drains and increasers-reducers. They are made of elastomeric PVC, and they are strong, resilient and unaffected by soil conditions. They are also resistant to chemicals, ultraviolet rays, fungus growth, and normal sewer gases due to the inert nature and physical properties of the material. And, they're leak-proof, root-proof and seal against infiltration and ex-filtration.

# Review Questions

1.  What are three types of pipe used in above ground water systems?

    _____

2.  What does PVC pipe look like when compared to CPVC?

    _____

3.  What is the thinnest copper pipe used in water distribution?

    _____

4.  What are flexible couplings used for?

    _____

5.  How long has CPVC been used for water systems?

    _____

# 7 Cast Iron Pipe and Other Types of Pipes

## Performance Objectives

After reading and studying this chapter, you will:

- Know when and where cast iron pipe is used and how to work with it.

- Know how to identify iron pipe.

- Understand the words used to relate to cast iron and its fittings.

- Know the advantages and disadvantages of cast iron being used for plumbing purposes.

- Know which tools to use to cut cast iron.

- Know how heavy the pipe is and the tools to use when working with it.

- Be able to perform a cast iron lead pouring.

- Learn how to pack oakum firmly in a joint.

- Understand why various other pipes are needed.

- Be able to answer the review questions at the end of the chapter.

# Working with Cast Iron Pipe

The hub and spigot cast iron fittings shown here are of the OLD design still found in older homes and commercial buildings (see Figure 7-1). Since they are frequently encountered, it is best to have the pertinent data at hand to make replacement or repair of the fittings easier. Cast iron pipe has been used for many years to provide drains and vents (see Figures 7-2 to 7-6). Cast iron is still used for these purposes. However, most cast iron has been replaced with plastic pipe (though some modern systems still use cast iron pipe). There are two types of cast iron.

■ The first type is a service weight with hubs. This pipe is called service weight cast iron, or *bell and spigot* cast iron.
■ The other type is a lighter weight pipe that does not have a hub.

Service weight cast iron is usually joined by using caulked lead joints at the hub connections. However, modern adapters can be used to avoid working with molten lead.

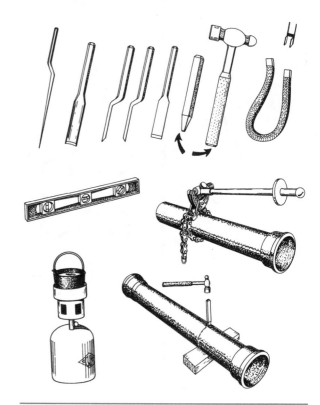

**Figure 7-2**  Tools used for working with cast iron soil pipe.

Lightweight, hub-less cast iron pipe is used in some jobs and is a much newer type of cast iron. The connections used for this type of pipe are a type of rubber coupling. Special couplings are designed for joining hub-less cast iron. They have a rubber band that slides over the pipe and fitting. Then a stainless steel band slides over the

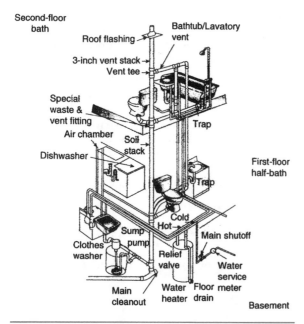

**Figure 7-1**  Note the use of cast iron pipe in the typical older home plumbing system. (Genova Products Inc.)

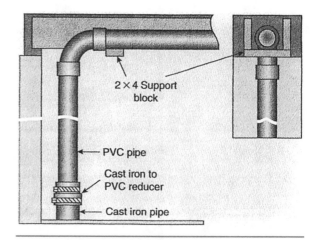

**Figure 7-3**  Support for a vertical pipe. (Renovation-Headquarters.com)

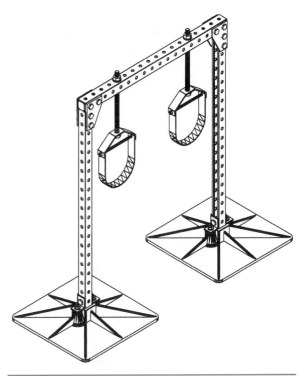

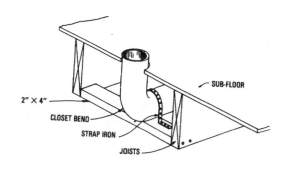

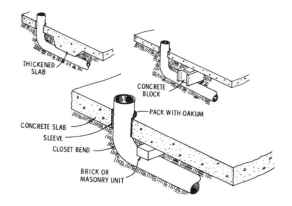

**Figure 7-4** Strapping horizontal run to a cross brace.

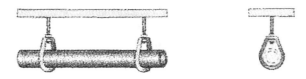

**Figure 7-5** Extra support is needed with no-hub. Note conventional hangers and supports in the ceiling to make the system rigid.

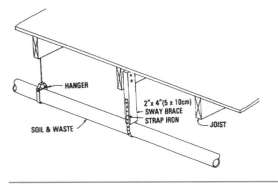

**Figure 7-6** Pipe through various types of floor slabs.

rubber coupling. Two clamps hold the band in place. Tightening the clamps on the band very tightly compresses the rubber coupling and makes the joint leak proof. Heavy rubber couplings work well with this pipe.

## Rubber Couplings

Heavy rubber couplings can be used with hubless cast iron. There are special rubber couplings designed to work with hub-less pipe. They can be used to match up cast-iron pipe to plastic pipe. However, an adapter is needed for the plastic pipe. The plastic adapter will have one end formed to the proper size to accommodate the

special coupling. Then the adapter is glued onto the plastic pipe and the special coupling slides over the factory-formed end of the adapter.

Working with cast iron pipe may require some special operations and tools. You may also need to be very careful and wear appropriate safety gear. You will be working with molten lead, open fires, and hot metals.

As mentioned earlier, cast iron pipe is heavy and hard to work. Perhaps the most exacting job in working with it is cutting it.

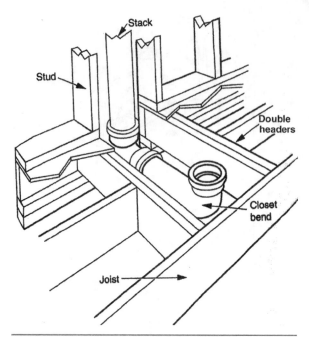

Figure 7-7    Bracing for a closet bend.

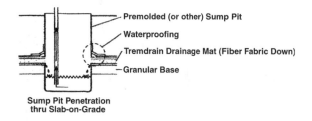

Figure 7-8    Slab-on-grade installation. (Tremco Inc.)

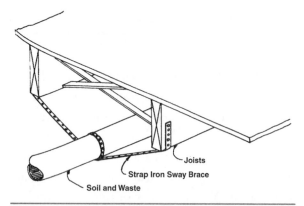

Figure 7-9    Horizontal pipe with sway brace.

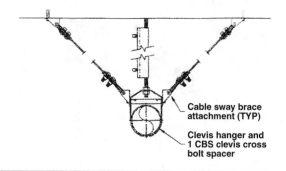

Figure 7-10    Sway brace. (NIBCO Inc.)

Before you cut cast iron pipe remember to measure it. Also remember to measure twice and cut once. Make a clearly visible mark all the way around the pipe where you want to cut. Make sure you allow the distance the pipe will jut into the fitting (see Figure 7-11).

You can use a hacksaw to cut the pipe, but that's a long and tedious job. We recommend that either a reciprocal saw or a chain cutter be used. Both can often be rented. Figure 7-12 shows a chain cutter in action.

If you have assistance, you can use a circular saw with a metal cutting blade. One person holds the saw and makes a shallow cut on the mark while the other person steadies the pipe and rotates it slightly as it is cut. Figure 7-13 illustrates this type of operation.

Another way to cut the pipe is to score it with a saw and finish the job by using a chisel and hammer to deepen the cut.

After the pipe has been cut it will be necessary to join them. One method is shown in Figure 7-14. The hub of the top pipe is fitted into the joint and then firmly tamped in place. You leave about ¼- to ½-inch between the packed oakum and the top of the flange. Next, lead is melted and poured into the gap between the flange and the side of the pipe.

Doing horizontal joints is a bit trickier because you simply can't pour the melted lead directly into the joint. You must use a gadget called a joint runner as seen in Figures 7-15 to 7-17. The joint runner forms a container around

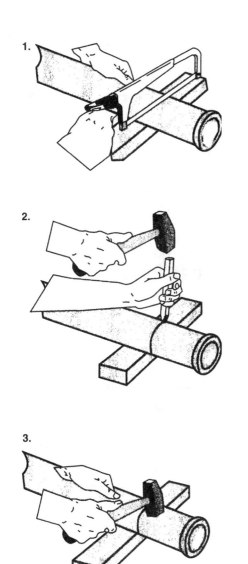

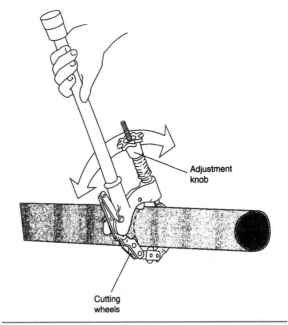

**Figure 7-12**   A chain cutter is a good way to cut cast iron pipe.

**Figure 7-11**   The old way to cut cast iron pipe still works. (1) Use a hacksaw to make a 1/16-inch deep cut around the pipe. (2) Then deepen the cut with a cold chisel. (3) Tap the end all around.

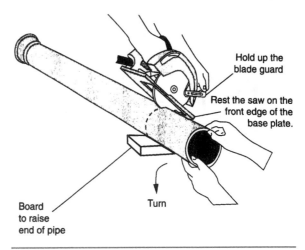

**Figure 7-13**   A circular saw with a metal cutting blade can be used to cut cast iron pipe.

the joint with a pouring hole at the top. You pour the melted lead into the hole until it fills up. Let the lead set, and expand it just as with a vertical joint.

Most codes today allow the use of flexible unions on horizontal and sloped runs. Refer to Figure 7-18. In fact, a variety of neoprene or rubber fittings are made for use with cast iron pipe, and you may be able to use them in your area. These include elbows (els), wyes, and tees. Check with your vendor and permit office first.

If you use them, you must cut off the flanges and hubs so that each pipe is just a straight cylinder. In some areas you may be able to buy cast iron pipe without the flanges and hubs.

Each fitting has a flanged opening just like a plastic or copper fitting. You seal the fitting with a clamp around each flange. You must be very careful to support the pipe. The soft fitting will deform and leak if you allow any weight to rest upon it.

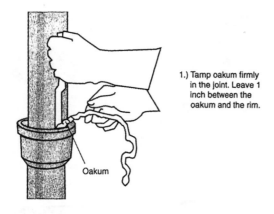

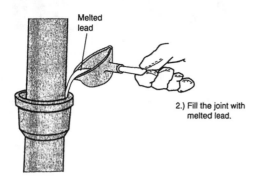

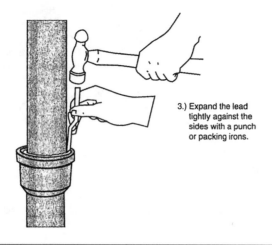

1.) Tamp oakum firmly in the joint. Leave 1 inch between the oakum and the rim.

Oakum

Melted lead

2.) Fill the joint with melted lead.

3.) Expand the lead tightly against the sides with a punch or packing irons.

**Figure 7-14**   Sealing a vertical cast iron pipe joint. (1) Tamp oakum firmly in the joint. Leave 1 inch between the oakum and the rim. (2) Fill the joint with melted lead. (3) Expand the lead tightly against the sides with a punch or packing irons.

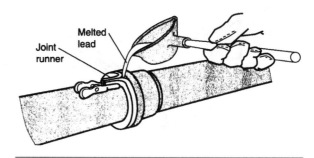

Melted lead

Joint runner

**Figure 7-15**   A joint runner is used to "lead" a horizontal cast iron pipe joint.

# Working with Clay and Fiber Pipes

Both clay and fiber pipes are only used for drain or sewer lines outside the home. Most codes require that they be used no closer to the home than 5 feet. These lines must also be sloped properly to allow waste matter to flow into the main sewer. If you put too much slope on the line, the water will run down too quickly and leave the solid matter stuck in the pipe, where it may harden and clog the drain. If you do not put enough slope on the pipe, neither the water nor the solids will drain effectively. This would result in a slow drain and the possibility that the fluids will back up into the house.

Most codes will require either $1/8$- or $1/4$-inch of slope per foot. The most common is probably $1/8$-inch per foot (see Figure 7-16). When you must lay a greater angle of slope, most codes now require that you use PVC or steel pipe. The standard inside diameter for all these drains is 4 inches.

Vitrified clay pipe is a ceramic product. The clay has been hardened by partial fusion of the

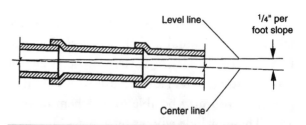

Level line

1/4" per foot slope

Center line

**Figure 7-16**   Clay and other drains should slope $1/4$-inch per foot.

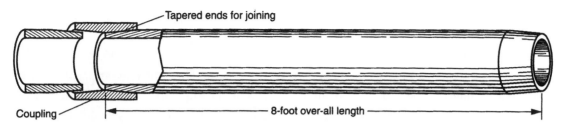

Figure 7-17  A joint runner is used to "lead" a horizontal cast iron pipe joint.

**4 in. ADS and Hancor to 4 in. Flexible PVC Coupling**

The Femco 4 in. Flexible PVC Coupling connects

**4 in. Ductile Iron and Asbestos Cement x 4 in. DWV**

This Femco 4 in. PVC Flexible Coupling

**4 in. Service Weight Cast Iron Hub x 4 in. Sch.**

The Femco 4 in. Flexible PVC Compression Coupling

**6 in. Clay x 4 in. DWV Flexible PVC Coupling**

The Femco 6 in. x 4 in. Elastomeric

**Oatey** 1-1/4 in. Chrome Semi Cast P-trap with Cleanout
Model# 507-1
NOT YET RATED

2 in. Cast-Iron Slip x Slip Adjustable P-trap
Model# K-6672-NA
NOT YET RATED

1-1/2 in. Chrome P-Trap without Cleanout
Model# 702-1
NOT YET RATED

1-1/2 in. Chrome P-Trap without Cleanout
Model# 2702-1
NOT YET RATED

6 in. EPDM Rubber Shielded Coupling
Model# PNH-66
NOT YET RATED

2 in. EPDM Rubber Shielded Coupling
Model# PHD-22

**Oatey** 1-1/2 in. Chrome P-Trap without Cleanout
Model# 704GBN-1
NOT YET RATED

3 in. Flexible PVC Coupling
Model# PHD-33
NOT YET RATED

Figure 7-18  Cast iron fittings.

clay particles by firing in a kiln. As a ceramic product, it is hard and brittle, but extremely durable and resistant to rot or corrosion. These pipes usually are purchased in two-foot lengths that have a wide flange on one end and a straight cylinder on the other, and are extremely durable and resistant to rot and corrosion.

Vitrified clay pipe is very difficult to cut. Probably the most accurate method is to use a circular power saw, as in Figure 7-13. You will, however, need a blade in the saw designed for cutting masonry rather than metal.

You can use a variety of tools to score (cut a shallow line) the mark where the pipe is to be cut. These could include a file, a hacksaw, a cold chisel, and even a hatchet. Necessity and lack of planning have often produced unusual solutions. However, you always want to cut the extra length off the straight cylindrical end.

Once the line is scored, you can tap around the line with a hammer to break the piece off the main pipe. It will usually break up in pieces rather than as a whole piece. Once you have all the pieces formed, you must lay the pipe in the trench at the proper slope and then cement the joints. Ordinary masonry cement works. It is often very tedious and time-consuming to get all the short lengths of pipe sloped properly.

Bituminous fiber pipe is a pipe made from rot-resistant fiber, such as fiberglass, and impregnated with bitumen, a form of tar. It is a lot easier to work with than clay pipe. Fiber pipe is very light in weight compared to clay pipe, but perhaps the best advantage is that it comes in lengths of 8 or 10 feet. The longer length makes it a lot easier to lay at the proper slope. But the ends of the pipe are tapered so that special fittings, such as in Figure 7-18, may be used to join the pieces. These fittings are sealed with a tar-based cement. Fiber pipe can be cut with ordinary wood-cutting tools such as a handsaw or a power saw with a regular woodcutting blade.

When you need to cut fiber pipe to a shorter length, probably the best and easiest way to join them is with a flexible sleeve coupling. You must cut the taper off the pipe to get a good seal, though.

## Other Pipe Types

There are a few other types of pipe that we haven't mentioned so far. These include brass pipe and chrome-plated brass or copper pipe. Brass pipe can be threaded or soldered. You work it just as you would copper pipe; you solder it, just as you do with galvanized pipe if you thread it.

Chrome-plated pipe shouldn't be soldered because the heat can discolor or damage the plating. Any type of plated pipe normally works with compression fittings, just like working soft copper tubing. In most cases, the fittings will also be chrome-plated and sized to fit the pipe you are working.

There are many plastic fittings available that can be used with soft copper and with various types of plastic lines. These are usually PB types of plastics with compression fittings. The advantage of plastic fittings is that they are softer and deform to make a seal with less force.

# Review Questions

1. What metal is used for cast iron pipe?

   _____

2. How do you cut cast iron?

   _____

3. What is meant by leading-in a joining of cast iron pipes?

   _____

4. Why are rubber couplings used in cast iron systems?

   _____

5. What are the three ways used to cut a cast iron pipe?

   _____

6. What is a hub-less cast iron pipe?

   _____

7. Why do some cast iron pipes need sway braces?

   _____

# 8 | Pumps

## Performance Objectives

After studying this chapter, you will:

- Be able to identify all the terms related to pumps.

- Know how to lubricate an electrically operated pump.

- Understand why pumps are needed in sewage systems.

- Be able to draw a diagram of a sewage system using a pump.

- Know how to check pumps for shorts and opens.

- Know how to choose a correct pump for a given system.

- Be able to answer the review questions at the end of the chapter.

## Some Pump Terms

- **Friction Head.** Friction head is the pressure expressed in psi or in the feet of liquid needed to overcome the resistance to the flow in the pipe and fittings.
- **Suction Lift.** Suction lift exists when the source of the supply is below the centerline of the pump.
- **Suction Head.** Suction head exists when the source of the supply is above the centerline of the pump.
- **Static Suction Lift.** Static suction lift is the vertical distance from the centerline of the pump down to the free level of the liquid source.
- **Static Suction Head.** Static suction head is the vertical distance from the centerline of the pump up to the free level of the liquid source.
- **Static Discharge Head.** Static discharge head is the vertical elevation from the centerline of the pump to the point of free discharge.
- **Dynamic Suction Lift.** Dynamic suction lift includes the sum of static suction lift, friction head loss, and velocity head.
- **Dynamic Suction Head.** Dynamic suction head includes static suction head minus the sum of friction head loss and velocity head.
- **Dynamic Discharge Head.** Dynamic discharge head includes the sum of static discharge head, friction head, and velocity head.
- **Total Dynamic Head.** Total dynamic head includes the sum of the dynamic discharge head plus the dynamic suction lift or discharge head minus dynamic suction head.
- **Velocity Head.** Velocity head is the head needed to accelerate the liquid.
- **Specific Gravity.** Specific gravity is the direct ratio of any liquid's weight to the weight of water at 62°F 62.4 lbs./cu. ft. or 8.33 lbs./gal.

- **Viscosity.** Viscosity is a property of a liquid that resists any force tending to produce flow. It is the evidence of cohesion between the particles of a fluid that causes a liquid to offer resistance analogous to friction. A change in the temperature may change the viscosity depending upon the liquid. Pipe friction loss increases as viscosity increases.
- **Static Pressure.** Static pressure is the water pressure required to fill the system.
- **Static System Pressure.** Static system pressure is the water pressure required to fill the system plus 5 psi.
- **Flow Pressure.** Flow pressure is the pressure the pump must develop to overcome the resistance created by the flow through the system.

## Installation and Clearance Requirements

1. The minimum recommended clearance around pumps is 24 inches. Maintain minimum clearance as required to open access and control doors on pumps for service, maintenance, and inspection.
2. Mechanical room locations and placement must take into account how pumps can be moved into and out of the building during initial installation and after construction for maintenance and repair and/or replacement.

## Lift Pumps or Sewage Pumps

Pumping stations are built when sewage must be raised to a point of higher elevation, or where the topography prevents downhill gravity flow. Special non-clogging pumps are available to handle raw sewage. They are insulated in structures called lift stations. Two basic types of lift station are dry well and wet well. A wet well installation has only one chamber or tank to receive and hold the sewage until it is pumped out. Specially designed submersible pumps and motors can be located at the bottom of the chamber, completely below the water level. Dry well installations have two separate chambers, one to receive the wastewater, and one to enclose and protect the pumps and controls. The protective dry chambers allow easy access for inspection and maintenance.

All sewage lift stations, whether wet well or dry well, should include at least two pumps. One pump can operate while the other is removed for repair. In this type of service, centrifugal pumps are required to pump either raw sewage or sludge. The solid precipitant that remains after the raw sewage has been treated chemically or bacterially is called sludge.

The chief difference between centrifugal pumps used for sewage and those used for sludge is the design of the impeller that is used. To avoid clogging, the impeller used for raw sewage is of an enclosed type, and it is usually wider, with two to four vanes, depending on the size of the pump.

The inlet portion of the vanes is usually rounded to reduce resistance to flow; it is shaped to prevent clogging by strings, rags, and paper, which tend to form a wad or ball of material.

The handling of sludge from a sewage treatment plant is more difficult than the handling of raw sewage, because larger quantities of the various solids are present. In addition to a properly designed impeller, the sludge pump is designed with a double-threaded screw in the inlet connection to force the sludge into the impeller. Each thread of the screw connects to an impeller vane. Solids or stringy materials that extend beyond the edges of the screw are cut up as they pass the edges of the screw and the flues of the screw housing.

When sewage pumps are installed they are usually relatively permanent, therefore the stationary pump parts (casing and base) should be substantial. The vertical type of sewage pump (see Figure 8-1) is designed for convenience in

inspection and maintenance. The cutaway view of the pump shown in Figure 8-2 shows how the motor is placed in a water-tight enclosure and the ball bearings are lubricated for life. The semi-axial impellers with large free passage ensure blockage-free operation at high efficiencies. A sewage pump used in lift station service is shown in Figure 8-3.

## Types of Pumps

There are a number of pumps that are made to very different specifications and for different jobs. Most of them are mentioned here, but most are identified by the work they were designed to do.

Pump selection should be based on a thorough understanding of the characteristics and fundamental principles of the basic types of pumps:

- Centrifugal
- Rotary
- Reciprocating

**Figure 8-1** Submersible propeller pump or large volumes of water or sewage. (ABS)

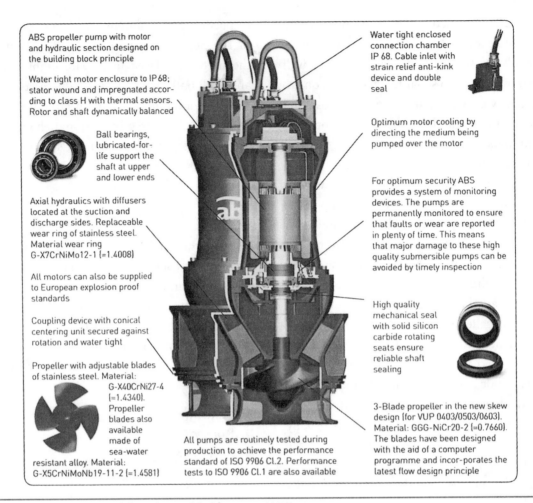

ABS propeller pump with motor and hydraulic section designed on the building block principle

Water tight motor enclosure to IP 68; stator wound and impregnated according to class H with thermal sensors. Rotor and shaft dynamically balanced

Ball bearings, lubricated-for-life support the shaft at upper and lower ends

Axial hydraulics with diffusers located at the suction and discharge sides. Replaceable wear ring of stainless steel. Material wear ring G-X7CrNiMo12-1 (=1.4008)

All motors can also be supplied to European explosion proof standards

Coupling device with conical centering unit secured against rotation and water tight

Propeller with adjustable blades of stainless steel. Material: G-X40CrNi27-4 (=1.4340). Propeller blades also available made of sea-water resistant alloy. Material: G-X5CrNiMoNb19-11-2 (=1.4581)

Water tight enclosed connection chamber IP 68. Cable inlet with strain relief anti-kink device and double seal

Optimum motor cooling by directing the medium being pumped over the motor

For optimum security ABS provides a system of monitoring devices. The pumps are permanently monitored to ensure that faults or wear are reported in plenty of time. This means that major damage to these high quality submersible pumps can be avoided by timely inspection

High quality mechanical seal with solid silicon carbide rotating seats ensure reliable shaft sealing

3-Blade propeller in the new skew design (for VUP 0403/0503/0603). Material: GGG-NiCr20-2 (=0.7660). The blades have been designed with the aid of a computer programme and incor-porates the latest flow design principle

All pumps are routinely tested during production to achieve the performance standard of ISO 9906 Cl.2. Performance tests to ISO 9906 Cl.1 are also available

**Figure 8-2**   Cutaway view of the submersible lift-pump. (ABS)

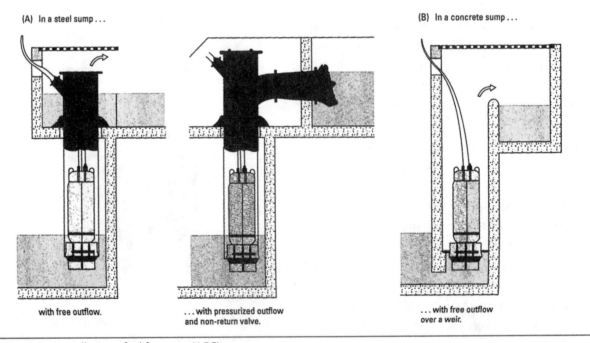

(A) In a steel sump . . .

(B) In a concrete sump . . .

with free outflow.

. . . with pressurized outflow and non-return valve.

. . . with free outflow over a weir.

**Figure 8-3**   Installation of a lift pump. (ABS)

The basic operating principles and various design characteristics or features adapt these pumps to specialized or unusual service conditions (see Figures 8-4 to 8-10).

The simplex or duplex reciprocating piston-type pumps are well adapted for the service requirements found in:

- Tanneries
- Sugar refineries
- Bleacheries

They are used for high-pressure service on water supply systems in country clubs, dairies, and industrial plants.

The general-purpose rotary gear-type pumps are designed to handle either thick or thin liquids, and they are designed to operate smoothly in either direction of rotation with equal efficiency. The thicker liquids such as:

- Roofing material
- Printing ink

- Fuel oil
- Gasoline and similar thin liquid

They can be handled by these pumps. The rotary gear-type pump is also adapted to:

- Pressure lubrication
- Hydraulic service
- Fuel supply
- General transfer work
- Pumping of clean liquids

The self-priming motor-mounted centrifugal pump has several applications, including:

- Lawn sprinklers
- Swimming pools
- Booster service
- Recirculation
- Irrigation
- Dewatering
- Sump and bilge
- Liquid fertilizers
- Chemical solutions

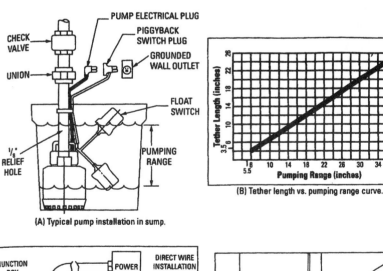

(A) Typical pump installation in sump.

(B) Tether length vs. pumping range curve.

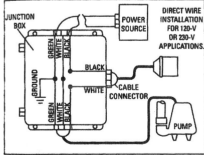

(C) Pumpmaster and pumpmaster plus hard wired.

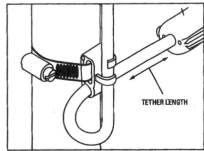

(D) Wide angle float mounting strap.

**Figure 8-4**   General specifications for sump pumps. (Gould)

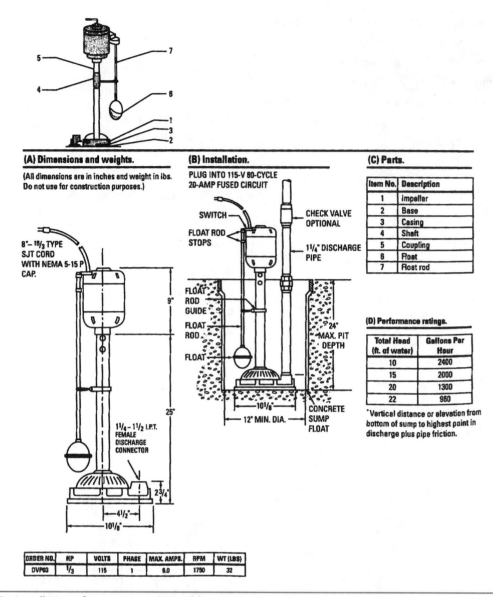

**(A) Dimensions and weights.**

(All dimensions are in inches and weight in lbs. Do not use for construction purposes.)

8"– 18/3 TYPE SJT CORD WITH NEMA 5-15 P CAP.

9"

25"

1¼ – 1½ I.P.T. FEMALE DISCHARGE CONNECTOR

2¾"

4½"

10⅛"

| ORDER NO. | HP | VOLTS | PHASE | MAX. AMPS. | RPM | WT (LBS) |
|-----------|-----|-------|-------|------------|------|----------|
| DVP03 | ⅓ | 115 | 1 | 6.0 | 1750 | 32 |

**(B) Installation.**

PLUG INTO 115-V 60-CYCLE 20-AMP FUSED CIRCUIT

SWITCH

FLOAT ROD STOPS

CHECK VALVE OPTIONAL

1¼" DISCHARGE PIPE

FLOAT ROD GUIDE

FLOAT ROD

FLOAT

24" MAX. PIT DEPTH

10⅛"

12" MIN. DIA.

CONCRETE SUMP FLOAT

**(C) Parts.**

| Item No. | Description |
|----------|-------------|
| 1 | Impeller |
| 2 | Base |
| 3 | Casing |
| 4 | Shaft |
| 5 | Coupling |
| 6 | Float |
| 7 | Float rod |

**(D) Performance ratings.**

| Total Head (ft. of water) | Gallons Per Hour |
|---------------------------|------------------|
| 10 | 2400 |
| 15 | 2000 |
| 20 | 1300 |
| 22 | 960 |

*Vertical distance or elevation from bottom of sump to highest point in discharge plus pipe friction.

**Figure 8-5**  Installation of a sump pump. (Gould)

Centrifugal pumps are also adapted to solids-handling applications such as:

- Sanitary waste
- Sewage lift stations
- Treatment plants
- Industrial waste
- General drainage
- Sump service
- Industrial process service
- Food processing
- Chemical plants

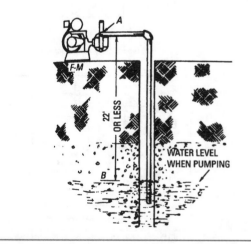

**Figure 8-6**  Shallow-well pump installation.

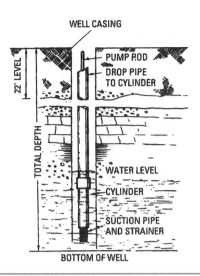

**Figure 8-7** Deep-well pump construction details.

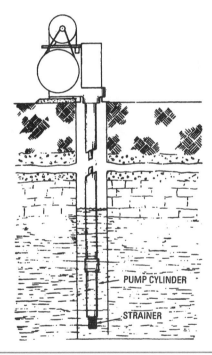

**Figure 8-8** Cylinder strainer placement for a deep-well pump.

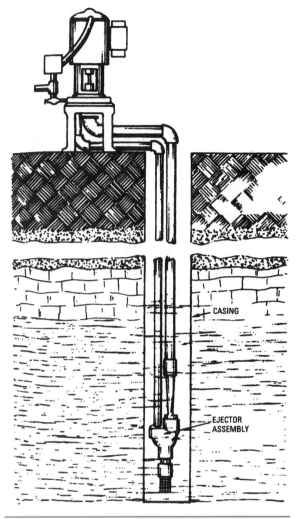

**Figure 8-9** A deep-well jet type pump installation.

The chemical and processing industries employ centrifugal pumps extensively in handling a wide variety of corrosive and abrasive liquids. The service life of pumps has been extended substantially by the use of rubber-lined pump parts for handling various liquids used in the chemical and paper industries. Pyrex or glass pumps are used to pump acids, milk, fruit juices, and other acid solutions. Therefore, the liquid being pumped is not contaminated by the chemical reactions between the liquid and the material in the pump parts.

Centrifugal pumps are also used to pump either raw sewage or sludge. As previously mentioned, the handling of sludge from a sewage treatment plant is more difficult than handling raw sewage, because larger quantities of various solids are present. The impeller is designed differently for handling raw sewage and sludge (depending on which is to be handled).

Magma pumps are designed to handle crude mixtures (especially of organic matter) that are in the form of a thin paste. For example, in the sugar-making process, heavy confectionery mixtures and non-liquids are involved. These pumps

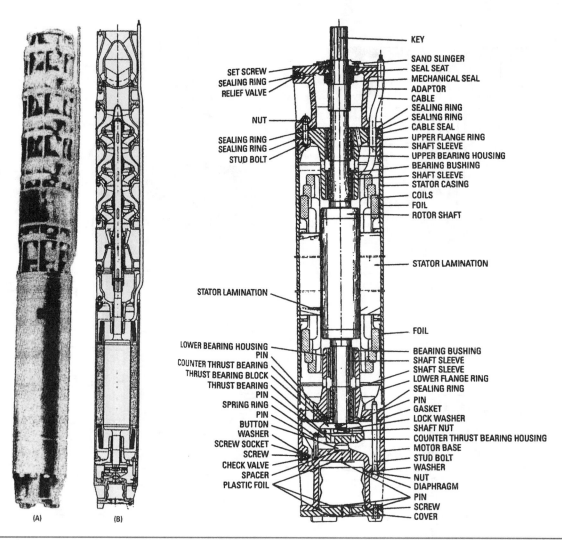

**Figure 8-10**    (A) Typical submersible pump motor with cross-sectional view. (B) Names of various parts of a submersible pump motor. (Plueger Submersible Pumps, Inc.)

are designed without inlet valves. The valves are not needed, because the liquid flows by gravity to the pump, and their function is performed by the piston of the reciprocating pump.

Other special service pumps include sump pumps, irrigation pumps, diaphragm-type pumps (which may be either closed diaphragm or open-diaphragm pumps), and shallow-well or deep well pumps (including the jet-type and submersible pumps). Turbine pumps and rubber impeller pumps are also finding wide use in industry and in marine applications.

# Review Questions

1. What does a lift pump do?

   _____

2. What does a magma pump do?

   _____

3. What are two types of deep well pumps?

   _____

4. What does Pyrex or glass pumps do?

   _____

5. What is a static discharge pump?

   _____

6. What are four special service pumps?

   _____

7. What are centrifugal pumps used for?

   _____

# 9 Toilets and Bidets

## Performance Objectives

After studying this chapter, you will:

- Be able to identify four types of toilets.

- Know the differences among the toilet types.

- Be able to explain how these different types of toilets operate.

- Recognize the bidet when you see one.

- Understand how the bidet can be a valuable item in a hospital.

- Be able to answer the review questions at the end of the chapter.

# Toilet Types

There are four types of toilet (or water closet) bowls:

- Blowout
- Reverse trap
- Siphon jet
- Wash-down

The toilet is the most important of all the sanitary fixtures. It is the most complicated of all the fixtures and the least understood. Remember this rule of thumb: the more water you see in the bowl, the better the closet. A large water surface tells you many things: the closet has positive siphon-jet action, it has a strong flushing action, and there is a deep water seal to guard against noxious gases. The toilet's construction, installation, and operation are important factors in determining the well-being and health of a building's occupants.

## Blowout

The blowout type of toilet (water closet) is operated with flush valves only (see Figure 9-1). The blowout bowl cannot be compared with any other type. It depends entirely upon a driving jet (D) action for its efficiency. It does not use siphonic action in the trap-way. It is rather economical in the use of water. Yet it has a large surface (A) that reduces fouling space. It also has a deep-water seal (B), and a large unrestricted trap-way (C). Blowout bowls are especially suitable for use in schools, offices, and public buildings. Besides, they are easy to clean under since they are above the floor and wall mounted.

## Reverse Trap

The reverse trap bowls are particularly suitable for installation with flush valves or low tanks (Figure 9-2). The flushing action and general appearance of the reverse-trap bowl are *similar to* the siphon jet, except that the water surface (A), depth of seal (B), and size of the trap-way (C) are smaller. That means less water is required for operation.

## Siphon Jet

The siphon jet bowl is a logical choice for the most exacting installation (see Figure 9-3). The flushing action of the siphon jet bowl is started by a jet (D) of water being directed through the up-leg of the trap-way. It instantaneously causes the trap to fill with water and the siphoning action begins. The quick, and relatively quiet action of the siphon jet bowl, when combined with its large water surface (A) and deep water seal (B) contribute to its general recognition by most people as the premier type of closet bowl.

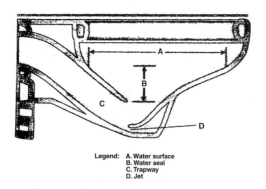

Legend:  A. Water surface
         B. Water seal
         C. Trapway
         D. Jet

**Figure 9-1**  The blowout water closet. (American Standard Brands)

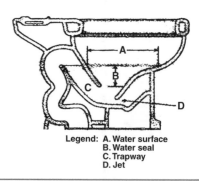

Legend:  A. Water surface
         B. Water seal
         C. Trapway
         D. Jet

**Figure 9-2**  The reverse-trap water closet. (American Standard Brands)

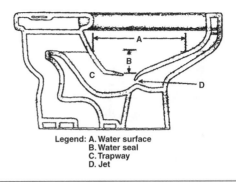

Legend: A. Water surface
B. Water seal
C. Trapway
D. Jet

**Figure 9-3** A siphon jet water closet.
(American Standard Brands)

## Wash-Down

The wash-down type bowl is simple in construction and yet highly efficient within its limitations. It will operate efficiently with a flush valve or low tank (see Figure 9-4). Proper functioning of the bowl is dependent upon siphonic-action in the trap-way accelerated by the force of water from the jet (D) directed over a dam. Wash-down bowls are widely used where low cost is the prime factor.

## Two-Piece Toilet Installation

Roughing-in dimensions for a vitreous china close-coupled combination toilet tank and bowl with siphon jet, whirlpool action, and elongated rim are shown in Figure 9-5. A typical siphon jet, wall-hung toilet bowl is shown in Figure 9-6 with dimensions. Figure 9-7 shows how a toilet works. Figure 9-8 illustrates the parts asso-

ciated with a flush tank mechanism. Knowing the proper and accepted name for each part is advantageous, especially when you are ordering replacement parts.

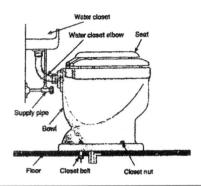

**Figure 9-5** Details of a floor-fastened water closet.
(American Standard Brands)

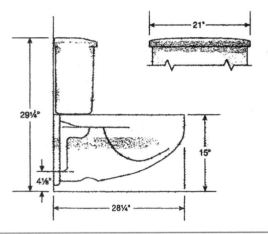

**Figure 9-6** Typical siphon jet, wall-hung toilet bowl.
(American Standard Brands)

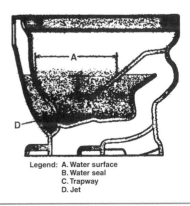

Legend: A. Water surface
B. Water seal
C. Trapway
D. Jet

**Figure 9-4** A wash-down water closet.
(American Standard Brands)

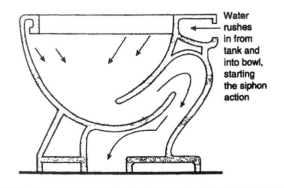

**Figure 9-7** Water flow in a flushed toilet.
(Genova, Inc.)

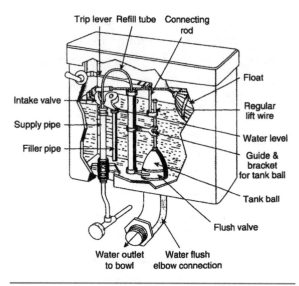

**Figure 9-8** Flush-tank mechanism for a toilet bowl. (American Standard Brands)

The rough-in dimensions for a siphon jet type of toilet tank and its bowl are given in Figure 9-9. Figure 9-10 shows how to locate the wax gasket on a 1.6-gallon flush toilet. The T-bolts sequence is shown in Figure 9-11, where connecting the water supply to the tank is shown in Figure 9-12.

## One-Piece Toilets

On the one-piece toilets (shown in Figure 9-13), the entire water closet is made as one piece so the tank and the seating area are cast as one piece. This arrangement has definite advantages and, of course, some disadvantages. As you can observe from the illustration, there are a number of one-piece toilets and this represents only one manufacturer: Eljer.

There are many types of toilets, both wall-mounted and floor mounted. See Figure 9-14. Again, this is representative of just one manufacturer.

Figure 9-15 shows the high rough-in water closet installation. This type is used with an automatic flush by Sloan Valve Company. Figure 9-16 illustrates how the wall hung water closet is mounted when used with an automatic flush-

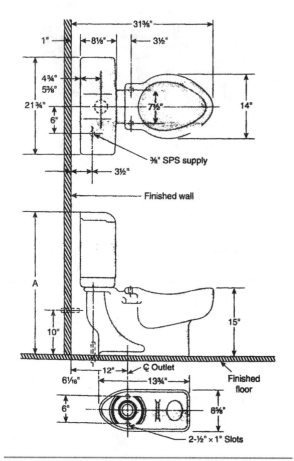

**Figure 9-9** Rough-in dimensions for a siphon type toilet tank and bowl. (American Standard Brands)

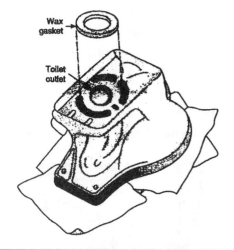

**Figure 9-10** Location of the wax gasket on a 1.6-gallon flush toilet. (American Standard Brands)

ing valve mounted properly. The other illustration shows an alternate ADA installation for the wall-hung bowl.

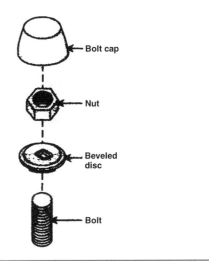

**Figure 9-11**  T-bolts sequence. (American Standard Brands)

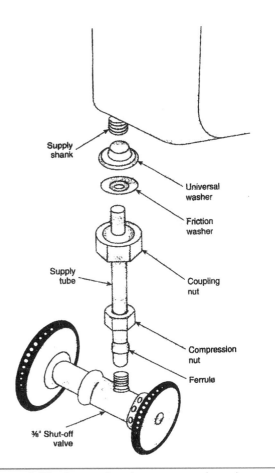

**Figure 9-12**  Connecting the water supply to the toilet tank. (American Standard Brands)

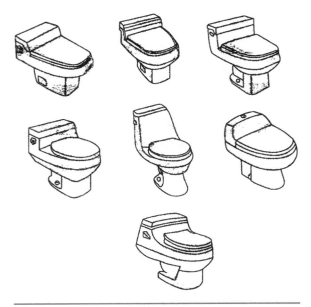

**Figure 9-13**  One manufacturer's offering of one-piece toilets. (American Standard Brands)

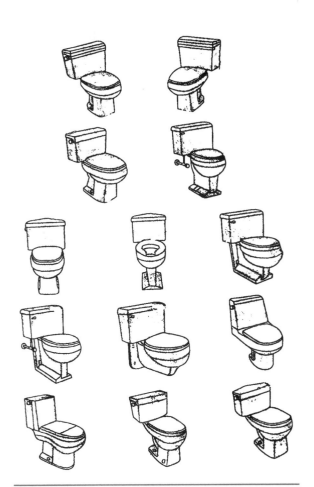

**Figure 9-14**  A variety of one manufacturer's toilet types. (American Standard Brands)

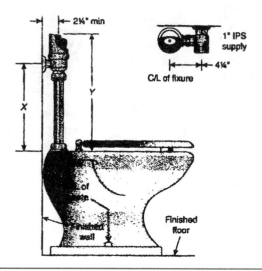

**Figure 9-15**   High rough-in water closet installation. (Sloan Valve Co.)

## Bidets

Bidets are a plumbing fixture located in the bathroom in some upper scale homes, but mostly in hospitals in special areas. They provide a soothing bath to the lower part of the body after a bowel movement or other activity. They are installed with hot and cold water outlets and with a mixing valve so they can furnish warm water. The lower part of the trunk is cleansed and soothed in the process. Those persons with hemorrhoids sometimes install a bidet for personal use. Most plumbers spend their career installing fixtures and never have the occasion to see one installed. They are capable of providing a *sitz* bath when needed without having to use a bathtub. Figure 9-17 shows a bidet.

**Figure 9-17**   Bidet

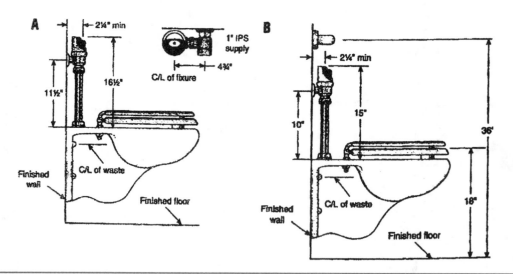

**Figure 9-16**   (A) Typical water closet installation. (B) Alternate ADA installation. (Sloan Valve Co.)

# Review Questions

1. Where is a blowout type of toilet used most frequently?

   _____

2. What is a reverse trap and where is it used?

   _____

3. How does the siphon jet water closet operate?

   _____

4. Where do you find the wash down type of toilet?

   _____

5. What is a bidet and where it is it used?

   _____

6. What is a sitz bath? Where is it used?

   _____

# 10 Urinals

## Performance Objectives

After studying this chapter, you will:

- Know where urinals are located.

- Know two ways to flush urinals.

- Understand how urinals are installed.

- Know where and how trough toilets are utilized.

- Know that there are two types of floor model urinals.

- Understand the amount of water needed per flush.

- Know what a wash-down pipe is and where it is used.

- Be able to answer the review questions at the end of the chapter.

# Urinals

Urinals are not included in most home bathroom installations. However, they are prevalent in such places as schools, restaurants, office buildings, and public rest rooms. There are two ways to flush these fixtures: automatically and manually.

## Manually

A typical toilet flush valve has been used for years on water closets and urinals to manually flush the bowls (see Figure 10-1). Most plumbing codes specify that no more than one fixture shall be served by a single flush valve. Each valve at each operation shall provide water in sufficient amount at a rate of delivery to completely flush the fixture and refill the fixture trap. Figure 10-1 is a typical toilet or urinal flush valve. Trough urinals shall have a flushing minimum capacity of 1½ gallons of water for each 2 feet of urinal trough length. Wash-down pipe shall be perforated so that the flush is made with an even curtain of water against the back of the unit. Trough urinals shall be not less than 6 inches deep with

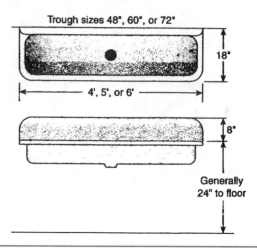

**Figure 10-2**   Rough-in dimensions for a trough type urinal. (Sloan Valve Company)

one-piece backs. Various state plumbing codes have fixture unit design basis tables and they do vary from locality to locality and state to state. For rough-in dimensions for a trough-type urinal, see Figure 10-2.

There are two different design types for floor mounted urinals—pedestal-type and stall-type (see Figures 10-3 and 10-4). Minimum dimensions are given in the figures.

Figure 10-5 shows a vitreous china siphon jet urinal with exposed valve and Figure 10-6 shows a typical wall-mounted urinal. Note: this one has an automatic flush valve mounted on top of it.

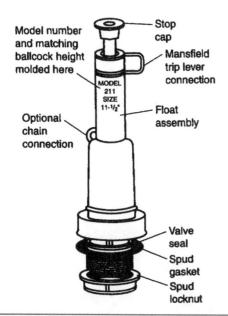

**Figure 10-1**   Typical toilet flush valve for water closet or urinal. (Mansfield Plumbing Products, Inc.)

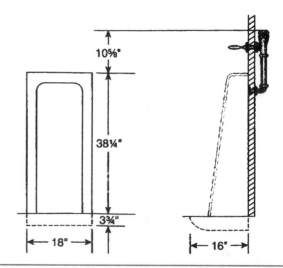

**Figure 10-3**   A stall-type urinal. (American Standard Brands)

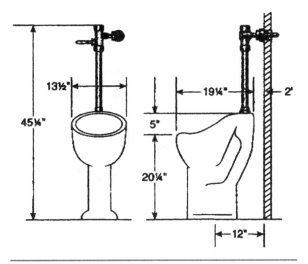

Figure 10-4   A pedestal-type urinal. (American Standard Brands)

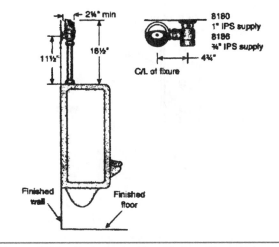

Figure 10-6   Typical urinal installation. (Sloan Valve Company)

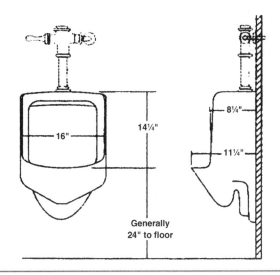

Figure 10-5   Vitreous china siphon jet urinal with exposed valve. (American Standard Brands)

## Automatically

Typical flush-o-meter valve installation (Figure 10-7) is covered in detail on the Internet. Complete step-by-step directions are given by the manufacturer. In this case, it is the Sloan Valve Company. Most low-consumption water closets and urinals call for 1.6 gallons (6-liters) per flush as a minimum with a flowing pressure of 25 psi (172 k.Pa). Most of the modern urinal installations require 1.0 gal/flush.

There are many installation possibilities for the wall mounted/floor models with the

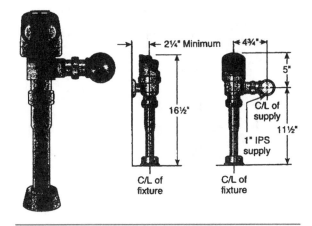

Figure 10-7   Battery operated, sensor-operated model water closet/urinal for floor mounted or wall-hung spud bowls. (Sloan Valve Company)

water source entering from the wall. Figure 10-8 shows the installing of the control stop with *bak-chek* or back check valve. Figure 10-9 shows the variations found in the capping-off of the water supply.

Figure 10-10 illustrates in the exploded view how the components of a flush-o-meter are assembled. Note how the flex tube diaphragm, O-ring, and regulator are installed in Figure 10-11. When installing a new regulator from a flex tube diaphragm kit, be sure to push the regulator past the O-ring when installing it. Note also that you never use more water than needed. Low-consumption water closets and urinals

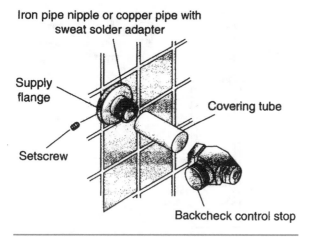

**Figure 10-8**   Control stop installation. (Sloan Valve Company)

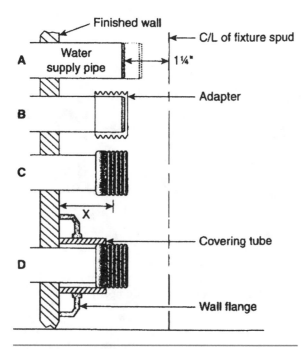

**Figure 10-9**   Various installation possibilities. (Sloan Valve Company)

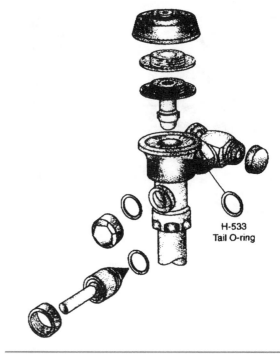

**Figure 10-10**   Removing components from existing flush-o-meter. (Sloan Valve Company)

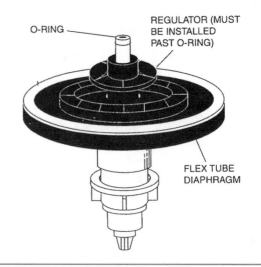

**Figure 10-11**   Flex tube diaphragm, O-ring, and regulator. (Sloan Valve Company)

will not function properly on excess water. The flush volume is controlled by the regulator flush kit. Table 10-1 shows the color of the regulator needed for the volume of the flush desired. It is important that the regulator be installed as shown in Figure 10-11.

**Table 10-1**   Regulator Colors

REGULATORS

The flush volume of the flex tube diaphragm kit is controlled by the regulator. Regulators are identified by color. Some flex tube diaphragm kits are supplied with multiple regulators. The installer must make sure the proper regulator is used when installing the flex tube diaphragm kit.

| REGULATOR (SOLD 6 PER PACKAGE) | | | |
|---|---|---|---|
| Code No. | Part No. | Description | Regulator Color |
| 5325122 | EBV-95 | Urinal-0.5 gpf/1.9 Lpf | Green |
| 5325122 | EBV-95 | Urinaf-1.0 gpt/3.8 Lpf | Green |
| 5325129 | EBV-102-2 | Urinal-1.5 gpf /5. 7 Lpf | Black |
| 5325130 | EBV-102-1 | Urinal-3.5 gpf/13~2 Lpf | White |
| 5325122 | EBV-95 | Closet-1.28 gpf/4.8 Lpf | Green |
| 5325122 | EBV-95 | Closet-1.6 gpf/6.0 Lpt | Green |
| 5325130 | EBV-102-1 | Closet-3.5 gpf/13.2 Lpf | White |
| 5325128 | EBV-101 | Closet-2.4 gpf/9~0 Lpf | Blue |

EBV-1020-A and EBV-1022-A are supplied with multiple flush volume regulators.
The installer must use the correct regulator when installing the kit.

Reproduced with permission of Sloan Valve Company.

# Review Questions

1. What is a urinal?

   _____

2. Why aren't urinals found in homes?

   _____

3. How is the color-code utilized in a urinal installation?

   _____

4. What is a flex-tube regulator used for?

   _____

5. What is an O-ring?

   _____

6. How is the O-ring used?

   _____

7. What is a back-check valve used for?

   _____

8. How much water is used for flushing a urinal?

   _____

9. What are the two ways to flush a urinal?

   _____

10. How many liters does it take to make a gallon?

   _____

# 11 Lavatories and Sinks

## Performance Objectives

After studying this chapter, you will:

- Know how to install a kitchen sink.

- Know the differences between hot and cold water service for a sink.

- Understand that the SPC code should be consulted before installation.

- Know the size of the water lines serving the sink.

- Understand the ADA guidelines regarding sinks.

- Be able to describe copper tubes and how they are sized.

- Be able to answer the review questions at the end of the chapter.

## Kitchen Sinks

Every home has a kitchen sink. It is an important fixture and installation must conform to the local code in order to properly serve its purpose.

The waste line should be 1½ inches (38 mm) according to the SPC (Standard Plumbing Code) and located 22¼ inches (565 mm) from the finished floor, 8 inches (203 mm) off the centerline of a double-compartment sink. A single-compartment sink should be roughed in at 25¼ inches (641 mm). Hot- and cold-water lines should be 23 inches (684 mm) from the finished floor—the hot should be 4 inches (102 mm) to the left of the centerline, and the cold 4 inches (102 mm) to the right.

If one compartment of a two-compartment sink is to be provided with a garbage disposal, the waste line should rough-in at 16 inches (406 mm) above the finished floor.

Figure 11-1 shows the installation of a sink that complies with the demands of the Americans with Disabilities Act (ADA) guidelines. This ensures that those sitting in a wheelchair can reach the sink.

## Service Sinks

Figure 11-2 shows the cross-sectional view of a mounted sink with a U-trap. The waste line or trap standard for these sinks is generally 3 inches (76 mm) SPC, and it roughs-in at 10½ inches (267 mm) above the finished floor. The water lines are generally roughed-in at 6 inches (152 mm) from the finished floor. The hot line is 4 inches (102 mm) to the left, and the cold line is 4 inches (102 mm) to the right. Figure 11-2 illustrates the installation of a sink with a U-trap. The faucet on a service sink may be sturdier and have a hook on the end to hold the bucket handle. In most instances, the service sink is mounted within a support framework (either a cabinet or steel support structure).

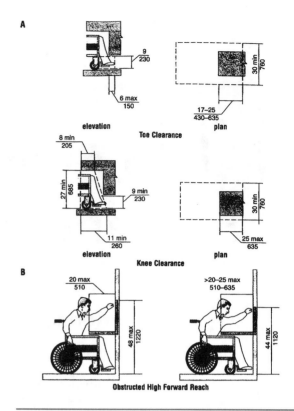

**Figure 11-1**   Americans with Disabilities guidelines for sink installation. (A) Toe and knee clearances. (B) Obstructed high forward reach ranges. (U.S. Access Board, *ADA Accessibility Guidelines for Buildings and Facilities* (ADAAG), www.access-board.gov/adaag/html/adaag.htm)

## Copper Tube Size

Some examples of minimum copper tube sizes for short-branch connections to these fixtures are shown in Table 11-1. The size of the short branches to individual fixtures can be determined by referring to Table 11-1. Generally, the mains servicing these fixture branches can be sized as follows:

- Up to three ⅜-inch (0.95 cm) branches can be served by a ½ inch (1.27 cm) main.
- Up to three ½-inch (1.27 cm) branches or up to five ⅜-inch (0.95 cm) branches can be served by a ¾-inch (1.91 cm) main.
- Up to three ¾-inch (1.91 cm) branches or correspondingly more ½-inch (1.27 cm) or ⅜-inch (0.95 cm) branches can be served by a 1-inch (2.54 cm) main.

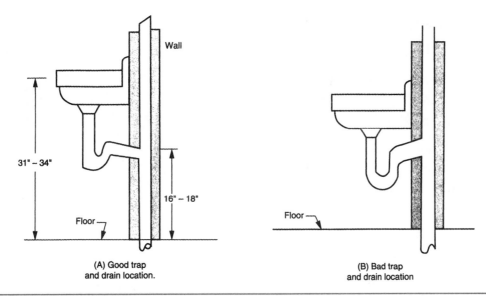

(A) Good trap
and drain location.

(B) Bad trap
and drain location

**Figure 11-2** Cross-sectional view of a mounted sink with a U-trap

**Table 11-1** Minimum Copper Tube Sizes for Short-Branch Connections to Fixtures

| Sizing of Copper Mains Servicing Individual Fixture Branches | | |
|---|---|---|
| **Main Size** | **Branches Serviced** | **Applicable Fixtures** |
| 1 in. | Up to three ³⁄₄-in. branches | Water closet, ≥ 1.6 gpf flush-o-meter valve |
| ³⁄₄ in. | Up to three ¹⁄₂-in. branches | Bathtub Combination bathtub/shower |
| ¹⁄₂ in. | Up to three ³⁄₈-in. branches | Bidet<br>Washing machine<br>Home dishwasher<br>Drinking fountain<br>Hose bibb<br>Lavatory<br>Kitchen sink<br>Laundry sink<br>Water closet, ≥ 1.6 gpf gravity tank<br>Water closet, 1.6 gpf flush-o-meter tank |

Data from the Copper Development Association, *The Copper Tube Handbook* (www.copper.org/publications/pub_list/pdf/copper_tube_handbook.pdf) and Bernaillo County Plumbing Code (www.bernco.gov/upload/images/forms/06_watersizingandcount.pdf).

# Water Demand

Each fixture in the system represents a certain demand for water. Some examples of approximate water demand in gallons per minute (gpm) or liters per minute (lpm) are found in Table 11-2.

**Table 11-2** Water Demand (Approx.)

| Water Demand | | Fixture |
|---|---|---|
| **gpm** | **Lpm** | |
| 0.75 | 2.84 | Drinking fountain |
| 2.0 | 7.57 | Lavatory faucet |
| 2.2 | 8.33 | Shower head |
| 2.5 | 9.46 | Self-closing lavatory faucet |
| 3.0 | 11.36 | Sink faucet<br>Water-closet tank ball cock |
| 3.5 | 13.25 | Flush valve (depends on design) |
| 4.0 | 15.14 | Bathtub faucet<br>Combination bathtub shower head<br>Laundry tub faucet |
| 5.0 | 18.93 | Hose bibb<br>Sill cock<br>Wall hydrant |

Data from the Copper Development Association, *Applications: Domestic Water Service and Distribution* (www.copper.org/applications/plumbing/apps/dom_h2o_distrn.html).

# Pressure Losses in a System Caused by Friction

Pressure available to move the water through a distribution system, or a part of it, is the main pressure minus the following:

■ The pressure loss in the meter
■ The pressure needed to lift water to the highest fixture (called the static pressure)
■ The pressure needed at the fixtures themselves

The remaining available pressure must be adequate to overcome pressure losses created by friction during flow of the total demand. In order to arrive at the total demand, it is necessary to know the intermittent and continuous flow-fixtures through which the distribution system furnishes its various parts. Tube sizes are selected in accordance with the pressure losses caused by this friction. Actually, the design operation may involve repeating the steps in the design process to readjust pressure, velocity, and size in order to achieve the proper balance between the combination of main pressure, tube size, velocity, and available pressure.

# Lavatories

There is an unlimited variety of lavatories for use in the home, school, business and industry. There are wall-hung lavatories, usually with two legs and attached to a wall. There are pedestal lavatories made of porcelain, and now some of stainless steel and other types of materials. See Figure 11-3 for some examples. Other types are self-rimming lavatories that drop into a hole usually cut in a countertop, as seen in Figures 11-4 and 11-5.

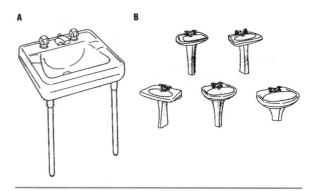

**Figure 11-3**   (A) Wall hung lavatory with legs. (B) Pedestal lavatories. (American Standard Brands)

**Figure 11-4**   Self-rimming lavatories. (American Standard Brands)

**Figure 11-5**   Self-rimming lavatories. (American Standard Brands)

Typical lavatory accessories include faucets and the various nuts and hardware that go with the installation of the faucets. One of the older antique types of faucets is shown in Figure 11-6. Figure 11-7 shows the lavatory plug top or

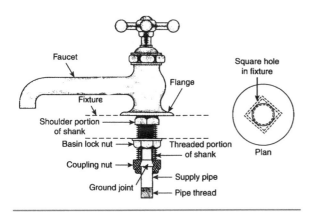

**Figure 11-6** Typical lavatory faucet accessories. (American Standard Brands)

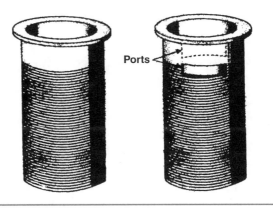

**Figure 11-7** Lavatory plug top or waste connection fittings. (American Standard Brands)

waste-connection fittings. There are two types: the plain one on the left and the one on the right, which is a ported fitting.

Every sink has a trip-lever mechanism. The one shown in Figure 11-8 shows a typical pop-up type.

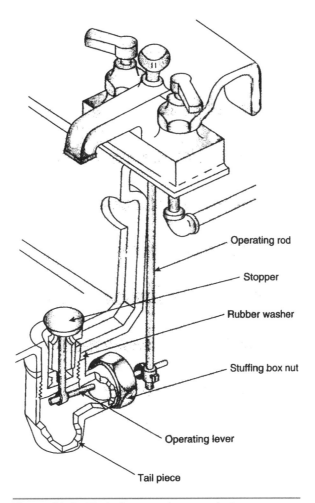

**Figure 11-8** Typical pop-up lavatory drain. (American Standard Brands)

# Review Questions

1. Why does a house need a kitchen sink?

   _____

2. What type of code is concerned with sinks?

   _____

3. What are two different types of kitchen sinks?

   _____

4. What is the meaning of the word lavatory?

   _____

5. What are three pressure losses due to friction?

   _____

6. What is a self-rimming lavatory?

   _____

7. What is the purpose of a U-trap?

   _____

# 12 Showers

## Performance Objectives

After studying this chapter, you will:

- Know how a shower head works.

- Understand how certain code requirements are necessary to protect the person using the shower.

- Know which federal energy standards are applicable to shower water usage.

- Be able to describe the processes employed by the Act-O-Matic shower head in order to be self-cleaning.

- Be able to answer the review questions at the end of the chapter.

## Plumbing Codes

Most plumbing codes state that all shower receptacles shall have watertight pans, except those built directly on the ground. When shower receptacles are required for ground surfaces, such receptors shall be built of dense, non-absorbent and non-corrosive materials with smooth impervious surfaces.

When shower pans are required for building installation, these pans shall be constructed of at least 4-pound sheet lead or 24-gauge copper with corners folded. The corners shall be soldered or brazed in an approved manner and insulated from the rest of the structure with at least 15-pound asphaltic type material.

## Shower Heads

The Sloan Act-O-Matic self-cleaning, wall-mounted shower head with adjustable spray direction is designed for institutional use. The chrome-plated shower head has the following features:

■ It is spring loaded and self-cleaning, with a spray disk that prevents particle clogging and a "cone-within-a-cone" spray pattern for total body coverage.
■ It has an adjustable spray angle and pressure compensating (2.5 gpm/9.4 lpm) flow control and is made of all-brass construction.
■ The mounting plate is vandal-resistant with its mounting screws.
■ The water inlet is ½-inch IPS pipe nipple with a shower head that is in conformance to all requirements of the ANSI/ASME Standard A 112.18.lM, CSA B-125 and the United States Federal Energy Policy Act (see Figure 12-1).

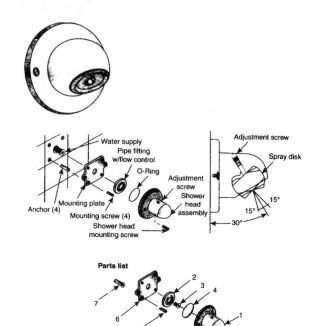

**Figure 12-1** Self-cleaning, wall-mounted shower head with adjustable spray direction and integral 2.5-gpm flow control. (Sloan Valve Company)

## Shower-Tub Diverters

Shower-tub diverters are similar to the one shown in Figure 12-2. A diverter functions in the same manner as a faucet. For stem-type diverters, turning the handle causes the stem to move into the valve seat and direct the water to the shower head. Figure 12-3 shows an exploded view of the tub and shower faucet. Figures 12-4 through 12-6 show exploded views of the shower head, shower bath valve and the cartridge type shower and bath valve.

## Shower Stalls

Take a look at Figure 12-7 for dimensions of shower stalls.

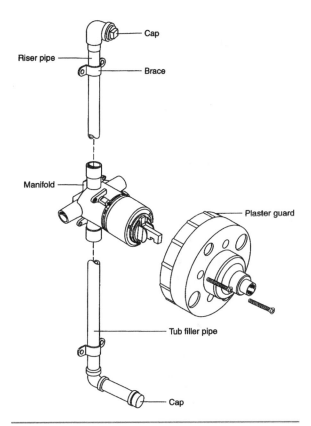

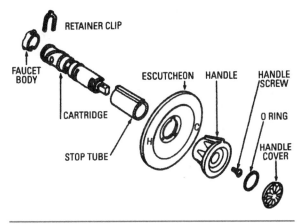

**Figure 12-4** Exploded view of cartridge-type shower and bath valves. (Plumb Shop)

**Figure 12-2** Tub-shower cutaway view of a diverter. (American Standard Brands)

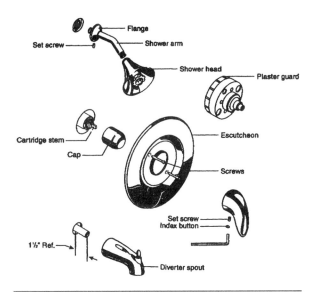

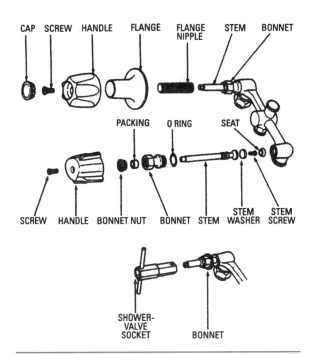

**Figure 12-5** Exploded view of shower and bath valve. (Plumb Shop)

**Figure 12-3** Exploded view of tub and shower faucet. (American Standard Brands)

## IMPORTANT

All plumbing is to be installed in accordance with applicable codes and regulations. Flush all water lines until the water is clear before installing the shower head. The water lines must be sized to provide an adequate flow of water for each shower head. Minimum water pressure of 15 psi (103 kPa) is required.

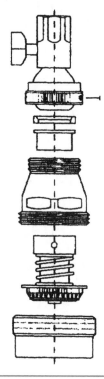

**Figure 12-6**  Exploded view of shower head. (Sloan Valve Company)

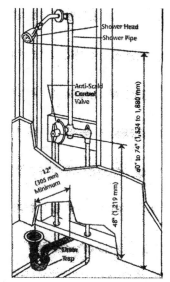

A.  Grab bars 33–36 in. above the floor should be provided on three walls of roll-in showers. For transfer-type shower stalls, the grab bar should extend across the control wall and back wall to point 18 in. from the control wall.

B.  There should be a ½ in. maximum threshold beveled with a slope not steeper than a 1:2 ratio.

C.  Minimum inside dimensions for transfer-type shower stalls: 36 in. x 36 in., with 36 in. x 48 in. minimum clear floor space for access. Minimum inside dimensions for roll-in showers: 30 in. x 60 in., with 36 in. x 60 in. minimum clear floor space for access.

**Figure 12-7**  Shower stalls. (National Kitchen & Bath Association, www.NKBA.org)

## Review Questions

1. How much water does a self-cleaning wall-mounted shower head allow?

   _____

2. What does a diverter do?

   _____

3. Which local code standards are used to check shower installations?

   _____

4. Why do most local codes specify that shower pans be made of a metal such as lead?

   _____

5. What is a shower stall?

   _____

# 13 | Copper Tubing, Plastic Pipe, Galvanized Pipe, and Their Fittings

## Performance Objectives

After reading and studying this chapter, you will:

- Be able to discuss the various characteristics of plastic piping used for plumbing.

- Be able to describe some of the limitations of each type of pipe used in plumbing.

- Know why we have codes for plumbers to follow as well as where you go to reach a certified inspector.

- Be able to use many of the terms found in the plumbing trade.

- Know how to identify various types of tubing and pipe.

- Know the various types of piping available to the plumber.

- Know where to look for suppliers of piping.

- Be able to answer the review questions at the end of the chapter.

# Plastic Pipe

Plastic drain, waste, and vent (DWV) piping has been approved by local and state codes for a number of years. They include the following:

- Building Officials Conference of America
- Southern Building Code Congress
- International Association of Plumbing and Mechanical Officials
- Federal Housing Administration (FHA)

The following are some types of plastic pipe:

- **Polyvinyl chloride (PVC-Type- 1).** Polyvinyl chloride is strong, rigid, and economical. It resists a wide range of acids and bases but may be damaged by some solvents and chlorinated hydrocarbons. The maximum service temperature is 140°F (60°C). PVC is better suited to pressure piping.
- **Acrylonitrile-butadiene-styrene (ABS).** Usage of ABS has almost doubled compared with PVC in DWV piping systems. However, it is limited to 160°F (71.1°C} water temperatures at lower pressures, which is considered adequate for DWV use.
- **Chlorinated polyvinyl chloride (CPVC).** This meets national standards for piping l 80°F (82.2°C) water at pressures of up to 100 psi (689 kPa). It can withstand 200°F (93.3°C) water temperature for limited periods. CPVC is similar to PVC in strength and overall chemical resistance.
- **Polyethylene (PE).** This is a flexible pipe for pressure systems. Like PVC, it cannot be used for hot-water systems.
- **Polybutylene (PB).** This is flexible and can be used for either hot-water or cold-water pressure systems. Since no method has been found to chemically bond PB, solvent-weld joints cannot be used. Compression type joints are used instead.

- **Polypropylene.** This is a very lightweight material suitable for lower-pressure applications up to 180°F (82.2°C). It is used widely for industrial and laboratory drainage acids, bases, and many solvents.
- **Kem-Temp polyvinylidene fluoride (PVDF).** This is a strong, tough, and abrasive resistant fluorocarbon material. It has excellent chemical resistance to most acids, bases, and organic solvents and is ideally suited for handling wet or dry chloride, bromine, and other halogens. It can be used in temperatures of up to 280°F (l38°C).
- **Fiberglass-reinforced plastic (FRP) epoxy.** This is a fiberglass-reinforced thermoset plastic with high strength and good chemical resistance up to 220°F (104.4°C).

## Plastic Piping Expansion

The following are some expansion characteristics of PVC pipe:

- PVC Type 1: 100 feet (30.5 meters) operating at 140°F (60°C) will expand approximately 2 inches (50.8 mm).
- CPVC-Polypropylene and PVDF at the same temperature will expand approximately 31 inches (82.55 mm).

## Applications

Plastic pressure piping for hot- and cold-water supply is now permitted in FHA-financed rehabilitation projects. Plastic pipe enjoys markets in:

- Natural gas distribution
- Rural potable water systems
- Crop irrigation
- Chemical processing

Almost 100 percent of all mobile homes and travel trailers have plastic pipe.

**Table 13-1** Set Times

| | Average Initial Set Schedule for WELD-ON® PVC/CPVC Solvent Cements | | | | |
|---|---|---|---|---|---|
| Temperature Range | Pipe Sizes ½" to 1¼" 20mm to 48mm | Pipe Sizes 1½" to 2" 50mm to 63mm | Pipe Sizes 2½" to 8" 75mm to 200mm | Pipe Sizes 10" to 15" 250mm to 380mm | Pipe Sizes 15"+ 380mm+ |
| 60°– 100°F/16° – 38°C | 2 minutes | 5 minutes | 30 minutes | 2 hours | 4 hours |
| 40°– 60°F/5° – 15°C | 5 minutes | 10 minutes | 2 hours | 8 hours | 16 hours |
| 0°– 40°F/–18° – 5°C | 10 minutes | 15 minutes | 12 hours | 24 hours | 48 hours |

These figures are estimates based on testing done under laboratory conditions. Field working conditions can vary significantly. This chart should be used as a general reference only.

Initial set schedule is the necessary time to allow before the joint can be carefully handled. In damp or humid weather, allow 50% more set time.

Set times are *not* cure times. Piping systems can be pressure tested *only* after full observance of cure times.

Reproduced with permission of IPS® Corporation.

Two types of plastic pipe and fittings are commonly used for drainage systems:

- PVC
- ABS

## Plastic Tubing Joints

Follow these steps when working with joints in plastic tubing:

1. Cut the tubing.
2. Test-fit the joint.
3. Apply primer (for PVC and CPVC only).
4. Apply cement.
5. Assemble the joint.
6. Allow the joint to set. See Table 13-1 for set times.

## Tube Cutting

Be sure to use the right primer and/or solvent. Priming is essential with PVC and CPVC. No priming is needed with ABS. The recommended practice for making solvent-cemented joints with PVC and ABS pipe and fitting follows:

1. Pipe should be cut square. Use a fine-tooth handsaw and a miter box.
2. Use a fine tooth power saw with a suitable guide.
3. Regular pipe cutters may be used.

4. Hand-held cutters resembling a pair of pliers or a hedge trimmer are available for smaller diameter PVC tubing.
5. Remove all burrs with a knife, file or sand paper.

Table 13-2 gives you PVC water ratings at room temperatures for Schedule 40 pipe as well as for CPVC.

Table 13-3 gives ABS water pressure ratings at room temperature 73°F (23°C) for Schedule 40.

**Table 13-2** PVC Water Pressure Ratings

| Nominal Size (in.) | Schedule 80 (psi) | Schedule 40 (pal) |
|---|---|---|
| ½ | 509 | 358 |
| ¾ | 413 | 289 |
| 1 | 378 | 270 |
| 1¼ | 312 | 221 |
| 1½ | 282 | 198 |
| 2 | 243 | 166 |
| 2½ | 255 | 182 |
| 3 | 225 | 158 |
| 4 | 194 | 133 |
| 5 | 173 | 117 |
| 6 | 167 | 106 |

Data from the Engineering Toolbox, PVC Pipes—Pressure Ratings (www.engineeringtoolbox.com/pvc-cpvc-pipes-pressures-d_796.html).

**Table 13-3**   ABS Water Pressure Ratings

| Temperature | Pressure (psi) | | | |
|---|---|---|---|---|
| | Class B Pipe | Class C Pipe | Class D Pipe | Class E Pipe |
| 68°F (20°C) | 87 | 131 | 174 | 218 |
| 86°F (30°C) | 75 | 113 | 152 | 189 |
| 104°F (40°C) | 65 | 94 | 129 | 160 |
| 122°F (50°C) | 52 | 80 | 104 | 131 |
| 140°F (60°C) | 41 | 61 | 80 | 102 |
| 158°F (70°C) | 29 | 41 | 52 | 70 |

Data from the Engineering Toolbox. ABS Pipes—Pressure Ratings (http://engineeringtoolbox.com/abs-pipes-pressure-ratings-d_1594.html)

## NOTE

Forty-eight hours is considered a safe period for the ABS piping system to be allowed to stand vented to the atmosphere before pressure testing. Shorter periods may be satisfactory for high temperatures, small sizes of pipe, quick drying cement, and tight dry-fit joints.

Table 13-4 shows how often the plastic pipe should be supported in horizontal runs.

The industry does not recommend threading of ABS or PVC Schedule 40 Plastic Pipe. Do not use common pipe dopes on threaded joints. Some pipe lubricants contain compounds that may soften the surface, which under compression can set up internal stress corrosion. If a

**Table 13-4**   Horizontal Runs

| PVC Pipe Support Spacing | | |
|---|---|---|
| | Maximum Span* (ft) | |
| Pipe Size (in.) | Schedule 40 PVC | Schedule 80 PVC |
| ½ | 6 | 4 |
| ¾ | 8 | 4 |
| 1 | 8 | 5 |
| 1¼ | 10 | 5 |
| 1½ | 10 | 5 |
| 2 | 10 | 6 |
| 3 | 12 | 7 |
| 4 | 14 | 8 |
| 6 | 17 | 10 |

*This data applies to temperatures of 100°F (38°C).
Data from the City of San Diego Public Utilities Department, Section 15020—Pipe Supports (www.sandiego.gov/mwwd/business/cwspecs/pdf/15020.pdf).

lubricant is believed necessary, you can use ordinary Vaseline or pipe tape.

## Copper Tubing and Pipe

Table 13-5 shows the dimensions for copper pipe, galvanized pipe, PVC and CPVC. Note the inside and outside dimensions of the pipe and the depth of the fitting.

**Rigid copper pipe.** Rigid copper pipe is usually soldered together. Some pieces are threaded, but they are special order. Fittings are soldered.

**Flexible copper tubing.** Soft copper is flexible and comes in coils. It is easy to bend, but

**Table 13-5**   Pipe Dimensions

| Pipe Type | Rated Size (in.) | Actual Inside Diameter (in.) | Outside Diameter (in.) | Fitting Depth (in.) |
|---|---|---|---|---|
| PVC (STR) | 1 | 1 1/16 | 1 3/8 | 3/4 |
| | 3/4 | 13/16 | 1 1/8 | 5/8 |
| | 1/2 | 5/8 | 7/8 | 1/2 |
| CPVC (STR) | 1 | 1 | 1 3/8 | 3/4 |
| | 3/4 | 3/4 | 1 | 5/8 |
| | 1/2 | 1/2 | 5/8 | 1/2 |
| Copper pipe (type M) | 1 | 1 1/16 | 1 3/16 | 15/16 |
| | 3/4 | 13/16 | 7/8 | 3/4 |
| | 1/2 | 9/16 | 5/8 | 1/2 |
| Galvanized steel | 1 | 1 1/16 | 1 1/4 | 11/16 |
| | 3/4 | 13/16 | 1 | 9/16 |
| | 1/2 | 9/16 | 3/4 | 1/2 |
| | 3/8 | 1/2 | 5/8 | 3/8 |
| | 1/4 | 3/8 | 1/2 | 3/8 |
| | 1/8 | 5/16 | 3/8 | 1/4 |

**Figure 13-1** Copper tubing bender. (RIGID)

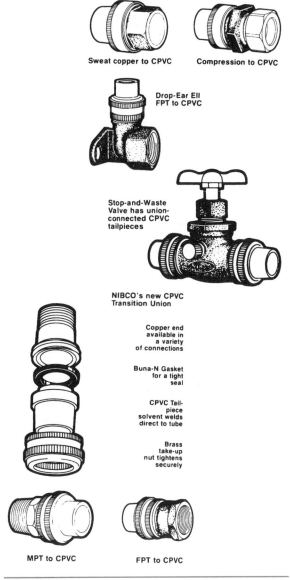

will collapse easily if not bent using a spring-like bending jig (see Figure 13-1). There are two types of fittings for soft copper tubing: compression fitting (Figure 13-2) and the flared fitting (Figure 13-3).

Tables 13-6 through 13-8 show details concerning copper tube and pipe.

## Galvanized Pipe

Table 13-5 shows the sizes and dimensions of galvanized pipe from ⅛-inch to a full inch rated size. This pipe is rigid and fittings are attached by way of threads which are cut in the pipe and the fitting. Remember to cut the pipe so the length includes the threaded end. Always remove the burrs from the cut pipe, both inside and outside the pipe. Use Teflon tape to seal the threads when putting a fitting on the end of the threaded pipe.

Figure 13-4 illustrates NIBCO's CPVC-to-metal unions and adapters. Figure 13-5 highlights NIBCO's drainage system adapters. Keep in mind that plastic pipe lighter than Schedule 80 should not be threaded. The wall thickness remaining after threading will not provide adequate strength. Figure 13-6 features crimp fittings for PB plastic tubing.

**Figure 13-2** Compression fitting with compression ring. (NIBCO)

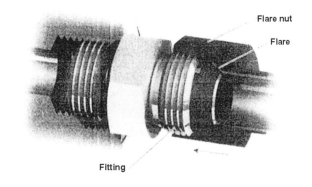

**Figure 13-3** Flared fitting. (Mathers Hydraulics Pty Ltd)

**Table 13-6**   Lengths of Copper Plumbing Tube

| Standard Copper Plumbing Tube Commercially Available Lengths | | |
|---|---|---|
| **Tube** | **Drawn** | **Annealed** |
| **Type K** | **Straight Lengths** | **Straight Lengths** |
| Available in diameters from ¼ inch to 12 inches (0.64 cm to 30.5 cm) | Up to 8 inches (20.3 cm) S.P.S. 20 ft (6.1 cm) 10 inches (25.4 cm) 18 ft (5.49 m) 12 inches (30.5 cm) 12 ft (3.66 m) | Up to 8 inches (20.3 cm) 20 ft (6.1 m) 10 inches (25.4 cm) 18 ft (5.49 m) 12 inches ( 30.5 cm) 12 ft (3.66 m) |
| | | **Coils** |
| | | Up to 1 inch (2.54 cm) S.P.S. 60 ft (18.29 m) 100 ft (30.48 m) 1¼ inches and 1½ inches (3.18 cm and 3.8 cm) 60 ft (18.29 m) 40 ft (12.19 m) 2 inches (5.1 cm) 45 ft (13.72 m) |
| **Type L** | **Straight Lengths** | **Straight Lengths** |
| Available in diameters from ¼ inch to 12 inches (0.64 cm to 30.5 cm) | Up to 10 inches (25.4 cm) S.P.S. 20 ft (6.1 m) 12 inches (30.5 cm) 18 ft (5.49) | Up to 10 inches (25.4 cm) 20 ft (6.1 m) 12 inches (30.5 cm) 18 ft (5.49 m) |
| | | **Coils** |
| | | Up to 1 inch (2.54 cm) 60 ft (18.29 m) 100 ft (30.48 m) 1¼ inches and 1½ inches (3.18 cm and 3.8 cm) 60 ft (18.29 m) 2 inches (5.1 cm) 40 ft (12.19 m) 45 ft (13.72 m) |
| **Type M** | **Straight Lengths** | **Straight Lengths** |
| Available in diameters from ⅜ inch to 12 inches (0.95 cm to 30.5 cm) | All diameters 20 ft (6.1 cm) | Up to 12 inches (30.5 cm) 20 ft (6.1 cm) |
| | | **Coils** |
| | | Up to 1 inch (2.54 cm) 60 ft (18.29 m) 100 ft (30.48 m) 1¼ inches and 1½ inches (3.18 cm and 3.8 cm) 60 ft (18.29 m) 2 inches (5.1 cm) 40 ft (12.19 m) 45 ft (13.72 m) |
| **DWV** | **Straight Lengths** | |
| Available in diameters from 1¼ inches to 8 inches (3.18 cm to 20.3 cm) | All diameters 20 ft (6.1m) | Not available |
| **ARC** | **Straight Lengths** | **Coils** |
| Available in diameters from ⅙ inch to 4½ inches (0.32 cm to 10.48 cm) | 20 ft (6.1 m) | 50 ft (15.2 m) |

**Table 13-7**   Properties of Copper Pipe

| Standard Copper Plumbing Tube | | |
|---|---|---|
| COMMERCIALLY AVAILABLE LENGTHS | | |
| **TUBE** | **DRAWN** | **ANNEALED** |
| Type K<br>Available in diameters from ¼ to 12" or .64 to 30.5 cm | **Straight Lengths:**<br>Up to 8" (20.3 cm) S.P.S.<br>  20 ft. (8.1 m)<br>10" (25.4 cm)<br>  18 ft. (5.49 m)<br>12" (30.5 cm)<br>  12 ft. (3.66 m) | **Straight Lengths:**<br>Up to 8" (20.3 cm)<br>  20 ft. (6.1 m)<br>10" (25.4 cm)<br>  18 ft. (5.49 m)<br>12" (30.5 cm)<br>  12 ft. (3.66 m)<br>**Coils:**<br>Up to 1" (2.54 cm) S.P.S.<br>  60 ft. (18.29 m)<br>  100 ft. (30.48 m)<br>¼ and 1½" (3.18 and 3.8 cm)<br>  60 ft. (18.29 m)<br>  40 ft. (12.19 m)<br>2" (5.1 cm)<br>  45 ft. (13.72 m) |
| Type L<br>Available in diameters from ¼ to 12" or .64 to 30.5 cm | **Straight Lengths:**<br>Up to 10" (25.4 cm) S.P.S.<br>  20 ft. (6.1 m)<br>12" (30.5 cm)<br>  18 ft. (5.49 m) | **Straight Lengths:**<br>Up to 10" (25.4 cm)<br>  20 ft. (6.1 m)<br>12" (30.5 cm)<br>  18 ft. (5.49 m)<br>**Coils:**<br>Up to 1" (2.54 cm)<br>  60 ft. (18.29 m)<br>  100 ft. (30.48 m)<br>1¼ and 1½" (3.18 and 3.8 cm)<br>  60 ft. (18.29 m}<br>2" (5.1 cm)<br>  40 ft. (12.19 m)<br>  45 ft. (13.72 m) |
| Type M<br>Available in diameters from ⅜ to 12" or .95 to 30.5 cm | **Straight lengths:**<br>All diameters<br>  20 ft. (6.1 m) | **Straight Lengths:**<br>Up to 12" (30.5 cm)<br>  20 ft. (6.1 m)<br>**Coils:**<br>Up to 1" (2.54 cm)<br>  60 ft. (18.29 m)<br>  100 ft. (30.48 m)<br>1¼ and 1½" (3.18 and 3.8 cm)<br>  60 ft. (18.29 m)<br>2" (5.1 cm)<br>  40 ft. (12.19 m)<br>  45 ft. (13.72 m) |
| DWV<br>Available in diameters from 1¼ to 8" or 3.18 to 20.3 cm | **Straight Lengths:**<br>All diameters<br>  20 ft. (6.1 m) | Not available |
| ACR<br>Available in diameters from ⅛ to 4⅛" or .32 to 10.48 cm | **Straight Lengths:**<br>  20 ft. (6.1 m) | **Coils:**<br>50 ft. (15.2 m) |

**Table 13-8**　Minimum Copper Tube Sizes for Short-Branch Connections to Fixtures

| Fixture | Copper Tube Size, Inches (cm) |
|---|---|
| Drinking Fountain | 3/8 (0.95) |
| Lavatory | 3/8 (0.95) |
| Water Closet (tank type) | 3/8 (0.95) |
| Bathtub | 1/2 (1.27) |
| Dishwasher (home) | 1/2 (1.27) |
| Kitchen Sink (home) | 1/2 (1.27) |
| Laundry Tray | 1/2 (1.27) |
| Service Sink | 1/2 (1.27) |
| Shower Head | 1/2 (1.27) |
| Sill Cock, Hose Bibb, Wall Hydrant | 1/2 (1.27) |
| Urinal (tank type) | 1/2 (1.27) |
| Washing Machine (home) | 1/2 (1.27) |
| Kitchen Sink (commercial) | 3/4 (1.91) |
| Urinal (flush valve) | 3/4 (1.91) |
| Water Closet (flush valve) | 1 (2.54) |

A. Two-Piece Design (CPVC Slip x Copper MIPT)

B. Two-Piece Design (CPVC Slip x Bronze FIPT)

C. Two-Piece Design (CPVC Slip x Bronze Solder)

D. Two-Piece Design (CPVC Slip x Brass Compression)

E. Drop-Ear Elbow

F. Bronze Stop and Waste Valve

**Figure 13-4**　CPVC to metal unions and adapters. (NIBCO)

A. Soil-Pipe Adapter
(Hub x No Hub)

B. Trap Adapter (Spg x SJ)

C. Soil-Pipe Adapter
(Hub x Spg)

D. Soil-Pipe Adapter
(Hub x Hub)

E. Adapter (Hub x FIPT)

F. Adapter (Hub x MIPT)

G. Adapter (Spg x FIPT)

H. Trap Adapter (Hub x SJ)

**Figure 13-5**   Drainage system adapters. (NIBCO)

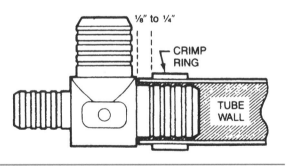

**Figure 13-6**   Crimp fittings for PB plastic tubing.
(US Brass)

# Review Questions

1. In what form is rigid copper tubing shipped to the plumber?

   _____

2. Why shouldn't you put threads on ABS or PVC plastic tubing?

   _____

3. How are rigid copper fittings fitted to the pipe?

   _____

4. What do the letters CPVC stand for?

   _____

5. What metal is used to coat steel piping to make it galvanized?

   _____

6. Where is Teflon tape used?

   _____

7. How long will galvanized pipe last in constant use?

   _____

8. Why should you always remove burrs from cut pipe?

   _____

9. What type of fittings are used for PB plastic tubing?

   _____

10. What does the NIBCO represent?

    _____

# 14 | Septic Tanks and Rural Water Systems

## Performance Objectives

After studying this chapter, you will:

- Know how septic tanks are produced and installed as part of a system.

- Know how to install a septic system.

- Understand how the drain system is made operational for a long time.

- Know how septic tank capacity is dependent upon the size of the house.

- Know the three materials most often used to make a septic tank.

- Understand how the drain fields operate and often malfunction.

- Know that cisterns are usually installed by specialists in that area.

- Know the difference between a well and a cistern.

- Be able to answer the review questions at the end of the chapter.

## Septic Tanks

Septic tanks operate for years without problems if properly installed (see Figure 14-1). The septic tank should be made water-tight and air-tight. This is so the bacterial action that disintegrates the solid matter can do its job (see Figure 14-2).

Remember that septic tanks are designed to operate full. While the bacterial action works on the solids, the liquid overflow drains off into the drain field. The tank does not need a cleanout, since the pre-cast slabs can easily be removed if it ever becomes necessary to clean the tank. Once the tank is cleaned, and the pre-cast slabs are set back in place, they can then be resealed with cement or mortar mix minus sand or gravel.

## Drain Fields

Drain fields can and will become spent in time, and new drain lines must be laid. These new lines must begin at the distribution box and extend out in new directions. Information may vary according to location and type of soil. Check your local code (see Figure 14-1).

*Examples*

*Hard Compact Soil*

- **2-bedroom house:** 750-gallon (2839 liter) tank with 200 feet (61 meters) of drain field.
- **3-bedroom house:** 1,000-gallon (3785 liter) tank with 300 feet (91.5 meters) of drain field.
- **4-bedroom house:** 1,000-gallon (3785 liter) tank with 400 feet (122 meters) of drain field.

With sandy soil, it may be allowable to install smaller tanks and less drain field footage.

The top of the tank is usually 6 inches (152.4 mm) to 10 inches (254 mm) below ground level.

The distance from the inlet invert (or bottom part of the inlet) to the septic tank from the top of the tank is generally 12 inches (305 mm). The outlet invert is generally 2 inches (51 mm) apart. Terra cotta or cement drain tile is used, measuring 4 inches (101.6 mm) inside and 12 inches (304.8 mm) long (see Figure 14-2).

## Septic Tank Drain Lines

Once the tank is set and leveled, the tank's drain lines are set. Begin by placing one drain tile between the outlet opening and the distribution box.

At each opening leading to a drain line, a ditch of about 24 inches (609.6 mm) wide is dug,

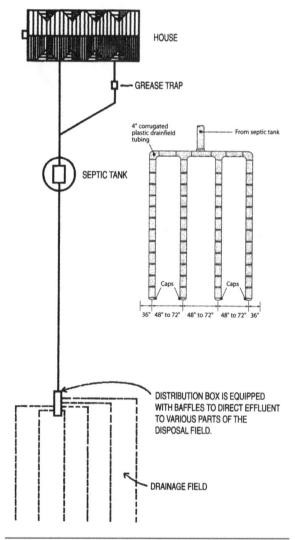

**Figure 14-1** Typical septic tank installation.

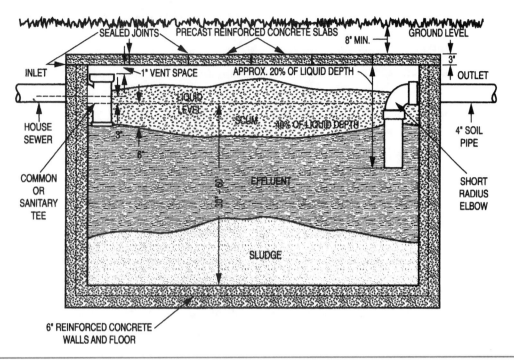

SEALED JOINTS PRECAST REINFORCED CONCRETE SLABS 8" MIN. GROUND LEVEL

INLET — 1" VENT SPACE APPROX. 20% OF LIQUID DEPTH OUTLET

HOUSE SEWER LIQUID LEVEL SCUM 40% OF LIQUID DEPTH 4" SOIL PIPE

COMMON OR SANITARY TEE EFFLUENT SHORT RADIUS ELBOW

SLUDGE

6" REINFORCED CONCRETE WALLS AND FLOOR

**Figure 14-2** Sectional of a typical septic tank.

6 inches (152.4 mm) below the outlet opening of the distribution box. Next, drive wooden pegs into the ground, beginning at each outlet of the box and spaced every 12 to 18 inches (304.8 to 457.2 mm) apart. The top of the peg should be leveled with the bottom of the outlet openings in the distribution box. Pegs should be laid level, or pitched approximately 1 inch (25.4 mm) every 100 feet (30.48 mm).

The crushed rock or gravel is installed to the level of the pegs or 6 inches (152.4 mm).

Now lay out the drain tile with care taken to space each tile approximately ⅜ inch (10 mm) apart. These cracks or spaces are to be covered with tar paper.

After the tile is set, each drain line will then receive more crushed rock until rock reaches 1 inch (25.4 mm) above the drain tile.

The last step is to cover the entire drain line with tar paper and then cover with earth (see Figure 14-3).

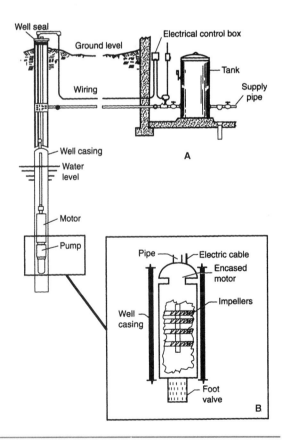

**Figure 14-3** Submersible pumps put motor and several impellers below water level.

## NOTE

A 1,000-gallon (3785-liter) septic tank contains approximately 134.75 cubic feet (3.787 cubic meters) of space. This size tank could measure 42 inches (1066.8) wide x 84 inches (2133.6 mm) long x 66 inches (1676.4) deep.

A 750-gallon (2839 liter) septic tank system contains approximately 100.25 cubic feet (2.838 cubic meters) of space. This size tank could measure 35 inches (889 mm) wide x 72 inches (1828.8 mm) long x 68.75 inches (1746.25 mm) deep.

*All measurements are taken from inside of the tank.

Septic tanks are usually made of concrete, but some are now made of fiberglass. The once-used metal tank is pretty much forbidden to be used in most codes. They rust and can cause trouble within a ten-year period of installation. Fiberglass tanks are easier to install and handle, but are not recommended for areas of heavy travel. The older idea of building the tanks of bricks and mortar and then coating the inside with asphaltum has now been abandoned. Most local building codes forbid their use.

Most plumbers, once they have determined that a problem is with the septic tank system, advise the owner to contact a certified Septic Tank Professional to perform the work.

## Rural Water Systems

The source of water for most households in America is a municipal water system, from which purified water is delivered under pressure to each home.

However, in outlying areas, water must be either extracted from private wells or from cisterns. People have been living with these systems for hundreds of years, but many are not familiar with how they work.

Wells are usually dug by professional well-drillers, with the plumbing handled by a plumbing contractor. See Figures 14-4 through 14-12 for a more detailed look at the rural water system components.

Cisterns are installed by specialized contractors, often in cooperation with a builder who installs the rainwater collection system, but once again a plumbing contractor usually installs the pumping and filtration system.

Sometimes to avoid the need to install purification equipment, homeowners elect to arrange for water to be delivered by truck. This works well only if the cistern is replenished before it runs dry, but the available source of water is limited.

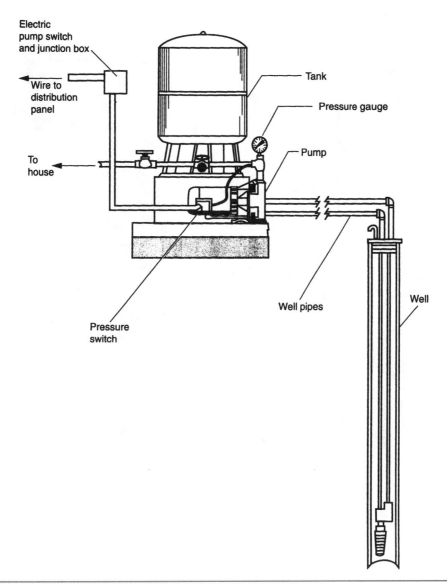

Electric
pump switch
and junction box

Wire to
distribution
panel

To
house

Tank

Pressure gauge

Pump

Pressure
switch

Well pipes

Well

**Figure 14-4** Typical well installation.

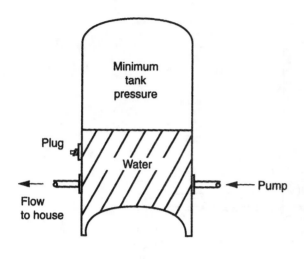

**Figure 14-5** Operation of air chamber tank. At minimum tank pressure, pump turns on.

**Figure 14-7** A gravity tank.

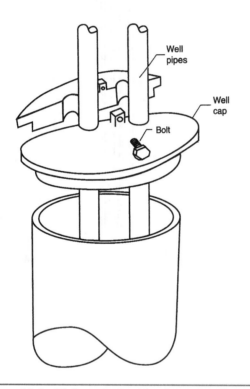

**Figure 14-6** Well cap is sized for the casing diameter.

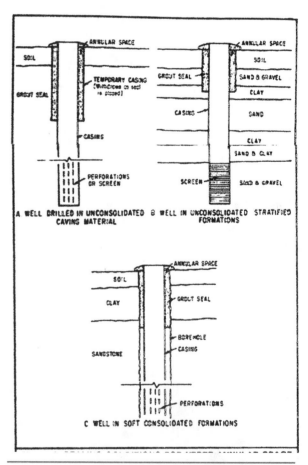

**Figure 14-8** Surface sealing requirements for water wells. (Courtesy of State of Texas)

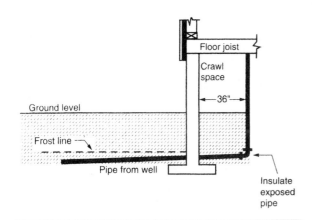

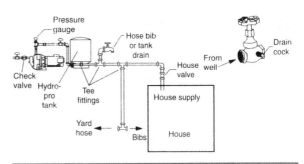

**Figure 14-9**   Pipe from well to home must be buried below the frost line.  Pump and tank can be protected in a basement, a well house, or by insulation.

**Figure 14-11**   Typical plumbing layout of supply system from well to house. A bleeder house valve with a drain cock will allow house pipe draining for winter closing.

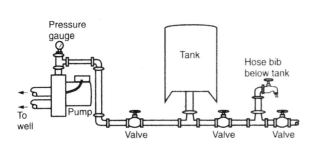

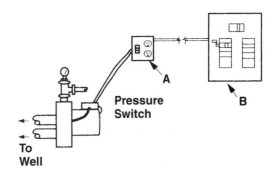

**Figure 14-10**   Desirable plumbing features at pump. Valves allow each section to be cut off for repairs or to drain and recharge tank. Hose bib functions as tank drain or area faucet. See also Table 14-1.

**Figure 14-12**   Desirable electrical features for a well pump. (A) Waterproof junction box with an on/off switch for pump and 110 Vac always "on" outlet. (B) Separate circuit breaker in main load center.

**Table 14-1**   Typical Pump Requirements for Siphon Jets and Packer Jet Systems

| Pump | Twin-Pipe Siphon Jet Systems | | | Packer Systems | | |
|---|---|---|---|---|---|---|
| | Min. Well Diameter | Control Valve Setting (psi) | Depth to Jet | Well Diameter | Control Valve Setting (psi) | Depth to Jet |
| ½ hp | 4" | 30 | 30-40 | 2" | 30 | 30-40 |
| | 4" | 30 | 30-60 | 2" | 30 | 30-60 |
| | 4" | 30 | 60-90 | 2" | 30 | 60-80 |
| | 4½" | 30 | 50-90 | 3" | 30 | 30-70 |
| | — | — | — | 3" | 30 | 50-80 |
| ¾ hp | 4" | 28 | 30-60 | 2" | 30 | 30-60 |
| | 4" | 29 | 70-90 | 2" | 32 | 60-90 |
| | 4" | 30 | 90-110 | — | — | — |
| | 4½" | 28 | 30-70 | 3" | 27 | 30-70 |
| | 4½" | 40 | 70-110 | 3" | 30 | 70-110 |
| 1 hp | 4" | 31 | 30-60 | 2" | 35 | 30-60 |
| | 4" | 33 | 60-80 | 2" | 36 | 60-100 |
| | 4" | 35 | 80-110 | — | — | — |
| | 4½" | 30 | 30-80 | 3" | 28 | 30-80 |

* Simple siphon pumps are used for depths to 20 feet.

# Review Questions

1. Why does a septic tank need to be charged with bacteria?

   _____

2. How would you describe a drain field?

   _____

3. How do you make a cistern?

   _____

4. Why should a septic tank be made air tight and water tight?

   _____

5. Who usually digs wells?

   _____

6. What is meant by the term "water collection system"?

   _____

7. What are septic tanks made of?

   _____

8. What is used as a filtering system?

   _____

9. Where do rural dwellers get their drinking water when the septic system fails?

   _____

10. Where do most Americans obtain the water they need to live in the city?

    _____

# 15 City Sewers

## Performance Objectives

After reading and studying this chapter, you will:

- Know why cities need a central sewage system.

- Be able to explain how the sewage system works.

- Understand which pipes can and cannot be used for sewage.

- Be able to describe how to assure a no-leak sewage system.

- Know where to use asbestos pipe.

- Know the difference between asbestos-cement and bituminous pipe.

- Be able to answer the review questions at the end of this chapter.

# City Sewers

Most cities are very protective of their sewers and sewage treatment plants. The health of a city can be affected adversely if the sewage system is not properly operated and protected from misuse. To accomplish this protection, a number of local ordinances have been used by cities to protect their inhabitants' health.

Sewer use regulations may include the following sections:

- Declaration of policy
- Use of public sewers required
- Private sewage disposal
- Building sewers and connection
- Use of the public sewers
- Restricted substances
- Fees
- Administration
- Enforcement
- Appeals
- Collection of costs and penalties
- Protection from damage
- Powers and authority of inspectors
- Penalties

The objectives are to:

- Prevent the discharge of pollutants into the city sewage system, which will interfere with the operation of the system or contaminate the resulting sludge.
- Prevent the discharge of pollutants into the city sewage system which, will pass through the system, inadequately treated, into receiving streams.
- Improve the opportunity to recycle and reclaim wastewater and sludge from the city sewage system.

Each county, state and local jurisdiction has its own rules and regulations for permits to discharge into the city sewers. It is best to navigate the various permitting agencies to obtain the proper permits for attaching to the city sewer system before starting to build on an "in city limits" lot.

A public sewer is sometimes referred to as a municipal sewer that is located in the street, alley or in an easement specified by the city in its approval of a plot plan for a subdivision.

Most public sewers consist of a 6-inch plastic pipe that is a lateral off the main sewer, which can be much larger than the laterals that run from it to various locations where building is planned for the near future. Figure 15-1 shows a plumbing drainage system leading up to the public sewer. The pipe sizes are identified in inches.

A certified plumber licensed to operate in any state, county or city is required to connect to or tap the city sewer line and connect the 4-inch pipe to the 6-inch sewer (see Figure 15-2).

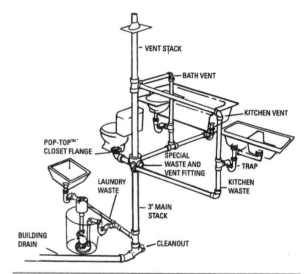

**Figure 15-1**   A typical drainage system for a house connected to a city sewer. (Genova Inc.)

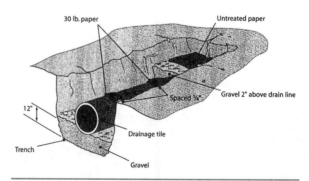

**Figure 15-2**   Sewer pipe load factors.

# Types of Sewer Pipes

## Asbestos-Cement Sewer Pipe

This type of pipe is usually restricted to exterior-buried use by local codes or environmental considerations.

## Bituminous Fiber Pipe

This is a very weak or fragile type of pipe no longer in general use, though it may be encountered in the sewer systems of older homes and commercial buildings. It is an inexpensive type of pipe and was once used as specified by government housing authorities (HUD).

## Cast Iron Pipe

Cast iron is coated with coal tar or centrifugal spun, or it may be extra-heavy cast iron depending on the expected or intended use of the system. If extra heavy is used, most codes require extra-heavy fittings for the interior of the building.

## Concrete Drain Pipe

Building sewers may also be made of concrete pipe and fittings. This type of pipe is susceptible to corrosion from acids and sewer gases and is not usually recommended. The vitrified clay pipe is more suitable for handling acids because of its glass-like finish.

## Copper Pipe

Copper is usually too expensive to use in the sewer system. Type DWV, KL and M are found in some older buildings.

## Vitrified Clay Pipe

VCP (vitrified clay pipe) is made from clay that has been subjected to vitrification.

Vitrification comes from the Latin meaning glass (vitrium). It is a process that fuses the clay particles to a very hard, inert, glass-like state. VCP is commonly used in sewer gravity collection mains because of its reasonable price and its resistance to almost all domestic and industrial sewage, particularly the sulfuric acid that is generated by hydrogen sulfide, a component of sewage. Only hydrofluoric acid and highly concentrated caustic wastes are known to attack VCP. Such wastes would not normally be permitted to be discharged into a municipal sewage collection system without adequate pre-treatment. You can also use standard or extra-strength vitrified clay pipes and fittings.

# Sewer Pipe Load Factors

For sewer pipe load factors see Tables 15-1 and 15-2. The illustration of the sewer pipe installations indicates the load factor on the pipe. Load factor is used in engineering the sewer system. Short load is a typical one and is provided by automobiles and truck traffic, road rollers and building foundations. Long loads include stacks of lumber, steel, and poles, and piles of sand, coal, gravel and so forth.

## Plastic Pipe

Building sewers may be made of clay pipe, ABS plastic or PVC schedule 40 or schedule 80 where the long load is the possible load on the buried pipe.

**Table 15-1**   Clay Pipe Size and Specifications

| Nominal I.D. (inches) | Nominal Laying Length (feet) | Approximate Weight (lbs/linear ft) | Minimum ASTM 3-Edge Bearing (lbs/linear ft) | Minimum ASTM High Strength* 3-Edge Bearing (lbs/linear ft) | Nominal O.D. Bell (inches) | Nominal Wall Thickness (inches) | Nominal O.D. Barrel (inches) |
|---|---|---|---|---|---|---|---|
| 4** | 6 | 10 | 2000 | — | $7\frac{3}{8}$ | $\frac{5}{8}$ | 5.1 |
| 6** | 6 | 20 | 2000 | — | $10\frac{3}{4}$ | $\frac{13}{16}$ | 8.0 |
| 8 | 6 | 30 | 2200 | — | 13.0/13.1 | 1 | 10.0/10.0 |
| 10 | 6 | 45 | 2400 | — | 16.3/16.3 | $1\frac{3}{16}$ | 12.9/12.9 |
| 12 | 6 | 60 | 2600 | — | 18.8/18.8 | $1\frac{3}{8}$ | 14.8/14.9 |
| BPC 15 | 7 | 90 | 2900 | 3400 | 22.6/22.6 | $1\frac{13}{16}$ | 18.5/18.6 |
| TEX 15 | 7 | 90 | 2900 | 3400 | 22.4/22.8 | $1\frac{13}{16}$ | 18.5/18.6 |
| 18 | $7\frac{1}{2}$ | 140 | 3300 | 4000 | 26.4/26.5 | $2\frac{1}{8}$ | 22.4/22.5 |
| 21 | 7 | 191 | 3850 | 4600 | 30.6/30.8 | $2\frac{5}{8}$ | 26.2/26.3 |
| 24 | 7 | 240 | 4400 | 5300 | 35.0/35.4 | $2\frac{15}{16}$ | 29.8/29.8 |
| 27 | 7 | 315 | 4700 | 5700 | 39.7/39.9 | $3\frac{1}{4}$ | 33.4/33.7 |
| 30 | 7 | 383 | 5000 | 6100 | 45.2/45.5 | $3\frac{7}{16}$ | 37.2/37.4 |
| 33 | 10 | 368 | 5500 | 6400 | 46.2/46.4 | $3\frac{1}{4}$ | 39.3/39.7 |
| 36 | 10 | 425 | 6000 | 6900 | 51.0/51.1 | $3\frac{1}{2}$ | 43.7/43.8 |
| 42 | $9\frac{1}{2}$ | 633 | 7000 | 7700 | 60 | $4\frac{1}{4}$ | 51.0 |

Dimension subject to normal variations as defined by ASTM C 700.

\* High strength pipe shall meet the requirements of ASTM C 700 and C 301 except that the minimum strength shall be shown in the table above.

\*\* Available in Band-Seal only.

## How to Specify JCP

All pipe shall be extra strength, vitrified clay, meeting the requirements of ASTM Specification C 700, as manufacture by Building Products Company LLC. Factory applied polyurethane joints conform to the material and performance standards of ASTM C 425 "Standard Specification for Compression Joins for Vitrified Clay Pipe and Fittings."

JCP Vitrified Clay Pipe and Band-Seal® Couplings meet or exceed the following standards:

■ ASTM C 700 and ASTM C 425

■ International Association of Plumbing and Mechanical Officials

| SPECIFICATIONS | |
|---|---|
| Vitrified Clay Pipe | ASTM C 700 |
| Compression Joint | ASTM C 425 |
| Clay Pipe Testing Method | ASTM C 301 |
| Installation Method | ASTM C 12 |

**Table 15-2**   Weight of Pipe-Trench Fill

| | Pressure Withstood by Standard-Weight Wrought-Iron Pipe | | | | |
|---|---|---|---|---|---|
| | **PIPE - SERVICE WEIGHT** | | | | |
| | | | Barrel | | |
| | Telescoping Length | Hub I.D. | O.D. | I.D. | Nominal Thickness |
| Size | Y | A | J | B | T |
| 2" | 2.50 | 2.94 | 2.30 | 1.96 | 0.17 |
| 3" | 2.75 | 3.94 | 3.30 | 2.96 | 0.17 |
| 4" | 3.00 | 4.94 | 4.30 | 3.94 | 0.18 |
| 5" | 3.00 | 5.94 | 5.30 | 4.94 | 0.18 |
| 6" | 3.00 | 6.94 | 6.30 | 5.94 | 0.18 |
| 8" | 3.50 | 9.25 | 8.38 | 7.94 | 0.23 |
| 10" | 3.50 | 11.38 | 10.50 | 9.94 | 0.28 |
| 12" | 4.24 | 13.50 | 15.88 | 15.16 | 0.36 |
| 15" | 4.25 | 16.95 | 15.88 | 15.16 | 0.36 |

Note: Length, inner diameter, and outer diameter are measured in inches.
Reproduced with permission of Charlotte Pipe and Foundry Company.

| | Pressure Withstood by Extra-Strong Wrought-Iron Pipe | | | | |
|---|---|---|---|---|---|
| | **PIPE - EXTRA-HEAVY WEIGHT** | | | | |
| | | | Barrel | | |
| | Telescoping Length | Hub I.D. | O.D. | I.D. | Nominal Thickness |
| Size | Y | A | J | B | T |
| 2" | 2.50 | 3.06 | 2.38 | 2.00 | 0.19 |
| 3" | 2.75 | 4.19 | 3.50 | 3.00 | 0.25 |
| 4" | 3.00 | 5.19 | 4.50 | 4.00 | 0.25 |
| 5" | 3.00 | 6.19 | 5.50 | 5.00 | 0.25 |
| 6" | 3.00 | 7.19 | 6.50 | 6.00 | 0.25 |
| 8" | 3.50 | 9.50 | 8.62 | 8.00 | 0.31 |
| 10" | 3.50 | 11.62 | 10.75 | 10.00 | 0.37 |
| 12" | 4.25 | 13.75 | 12.75 | 12.00 | 0.37 |
| 15" | 4.25 | 16.95 | 15.88 | 15.00 | 0.44 |

Note: Length, inner diameter, and outer diameter are measured in inches.
Reproduced with permission of Charlotte Pipe and Foundry Company.

# Review Questions

1. What size is the sewer pipe used in the street of a sub-division?

   _____

2. What is the difference between bituminous and asbestos pipe?

   _____

3. What is a load factor in terms of plastic sewer pipe?

   _____

4. What types of pipe can a builder use in a new sewer?

   _____

5. What type of plastic pipe may be used in buried sewer lines?

   _____

6. Why is VCP pipe used in sewers?

   _____

7. What is the Latin word for glass?

   _____

8. Where can you use vitrified clay pipe?

   _____

9. What material is used to make vitrified clay pipe?

   _____

10. What does HUD have to do with house building?

    _____

# 16 | City Drinking Water Systems

## Performance Objectives

After studying this chapter, you will:

- Understand how a city water service is installed and checked for proper installation.

- Know where the EPA is located.

- Know what EPA stands for.

- Know what MUD is and what they do.

- Be able to describe why a municipal utility district is needed in some areas.

- Know many of the contaminants of drinking water.

- Be able to take steps to actively protect a city's water supply.

- Know how to test drinking water.

- Know where to go for information on private or city water supplies.

- Be able to answer the review questions at the end of the chapter.

## Drinking Water

National Primary Drinking Water standards are legally enforceable standards that apply to public water systems. Primary standards protect public health by limiting the levels of contaminants in drinking water. The Environmental Protection Agency (EPA) website lists the contaminants, the potential health effects from ingesting the water, and sources of these contaminants in drinking water (www.epa.gov/ground-water-and-drinking-water/national-primary-drinking-water-regulations).

Municipal Utility Districts, city water departments, and other sources of public drinking water are required to list the quality of the water being sent to customers at least once a year. These usually arrive with the bill for the month. The contaminants listed on the EPA website will probably show up in the report. Using that information will allow you to protect yourself from any contaminated sources.

## Drinking Water Regulations

National Secondary Drinking Water Regulations (or secondary standards) are non-enforceable guidelines regulating contaminants that may cause cosmetic effects such as skin or tooth discoloration, or aesthetic effects such as taste, odor, or color in drinking water. The EPA recommends secondary standards for water systems, but does not require systems to comply. However, some states choose to adopt them as enforceable standards.

One contaminant with some serious effects is MTBE (methyl-t-butyl ether), found in some drinking water sources. All sources of drinking water contain some naturally occurring contaminants. As water flows in streams, sits in lakes, and filters through layers of soil and rock, it dissolves or absorbs the substances that it touches. According to its exposure, water transforms in composition and in physical parameters.

The Safe Drinking Water Act (SDWA) requires the EPA to establish and enforce standards that public drinking water systems must follow. The EPA delegates primary enforcement responsibility (all called primacy) for public water systems to states and Indian Tribes if they meet certain requirements.

Over 155,000 public water systems provide drinking water to most Americans. Customers that are served by public water can contact their local supplier and ask for information on contaminants in their drinking water, and are encouraged to request a copy of their Consumer Confidence Report. This report lists the levels of contaminants that have been detected in the water, including those by the EPA and whether the system meets state and EPA drinking water standards.

About 10 percent of people in the United States rely on water from private wells. Private wells are not regulated under the SDWA. People who use private wells need to take precautions to ensure that their drinking water is safe.

More details on the requirements are available from the EPA and can be found online. The following information is also found on the internet. It is placed here to help you to become aware of the problems associated with obtaining and providing drinking water for the millions of things it is used for. Water contaminants come from many locations. A few of them are listed here for your benefit.

## Public Water Supplies

Water systems can be public or private. A private water system is the responsibility of the homeowner. It is installed according to the latest rules and regulations furnished by the local health department.

The water main is thought of as having always been there. It has been, of course, if you live within a city's limits, or if you live in a Municipal Utility District (also known as MUD

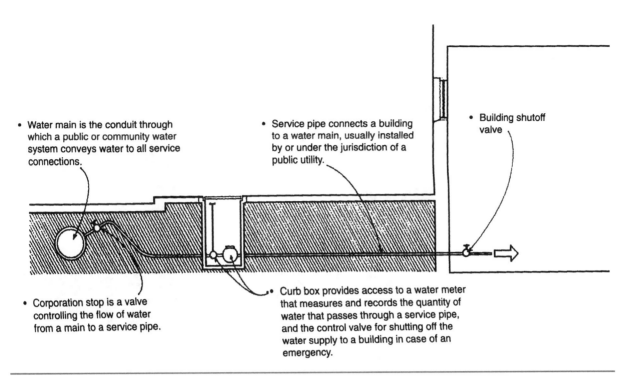

• Water main is the conduit through which a public or community water system conveys water to all service connections.

• Service pipe connects a building to a water main, usually installed by or under the jurisdiction of a public utility.

• Building shutoff valve

• Corporation stop is a valve controlling the flow of water from a main to a service pipe.

• Curb box provides access to a water meter that measures and records the quantity of water that passes through a service pipe, and the control valve for shutting off the water supply to a building in case of an emergency.

**Figure 16-1** A tapped city water main.

in some states). The water main is the conduit or large pipe through which a public or community water system conveys water to all service connections. Figure 16-1 shows how the typical home installation is made. The water main is usually located in the street—in front of the house—having been located there before the building permits were issued.

Once tapped, the water main furnishes water to the house by way of builder-installed plumbing, usually nothing more than a service pipe. This plumbing involves a *stop-valve*, or a place to turn off the water to the house. From there it goes to the curb box, which provides access to a water meter that measures and records the quantity of water that is used. This is done through a service pipe. There is usually a control valve for shutting off the water supply to a building

in case of emergency. The service pipe connects a house to a water main. The water main is usually installed by or under the jurisdiction of a public utility (see Figure 16-2).

After the service has been extended inside the building, a shut-off valve is installed. From there, all the water in the house is controlled by the valve's on/off function.

The EPA has a booklet to furnish information to those using private drinking water wells and other water sources. See Figure 16-3 or contact the EPA:

Environmental Protection Agency (EPA)
Ariel Rios Building
1200 Pennsylvania Avenue, NW
Washington, DC 20460
(202) 272-6167

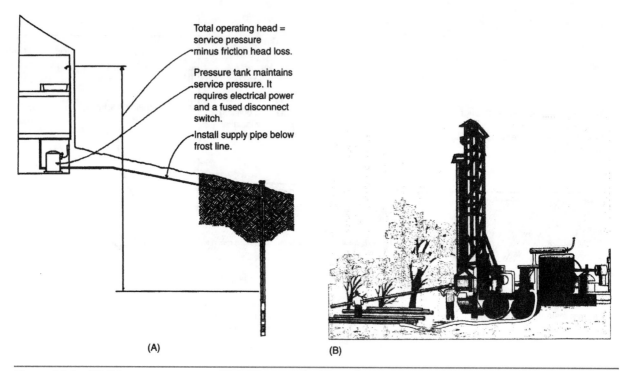

**Figure 16-2**    (A) Private water system. (B) Drilling a well with modern equipment.

**Figure 16-3**    The EPA has basic information about your water supply. You can obtain the information booklet by visiting the EPA website.

# Review Questions

1. Where does the city water come from?

   _____

2. The EPA recommends secondary standards for water systems.

   _____

3. Aluminum is one of the contaminants sometimes found in _____ water.

   _____

4. The National Primary Drinking Water Standards are:

   _____

5. What is another name for Municipal Utility Districts?

   _____

6. Where is the city water main located?

   _____

7. Where do you install a stop valve?

   _____

8. Where do water contaminants come from?

   _____

9. What does EPA stand for?

   _____

10. Where is the EPA located?

   _____

# 17

# Abbreviations and Symbols

## Performance Objectives

After studying this chapter, you will:

- Be able to identify symbols used on blueprints.

- Understand how to read the symbols on a blueprint.

- Recognize various abbreviations used in the plumbing field.

- Recognize all the present-day symbols used in mathematics.

- Be able to describe how pipe valves and fittings are used in plumbing a house.

- Be able to answer the review questions at the end of the chapter.

# Symbols

Plumbing symbols are used in drawing up plans for building houses, barns, commercial units, schools and almost anything with a roof on it. When making these drawings on a page of any size, it is necessary to reduce the details to fit within the limits of the paper. The working drawings for a building are complex. Items such as doors, windows, and electrical fixtures cannot be drawn on the plan in detail. They are represented by symbols so the drawing is easier to make and read. Figure 17-1 is one of those drawings with representative items as they would appear on a blueprint.

The American Standard Graphical Symbols for Piping are shown in Tables 17-1, 17-2, and 17-3.

This generally accepted page of symbols is used in the construction industry to allow quick and easy instructions to be prepared and read and put into use by plumbers on the job.

Figures 17-1 and 17-2 represent the standardized graphical symbols for pipe fittings and valves. As you can see, drafting or mechanical drawing is a form of shorthand.

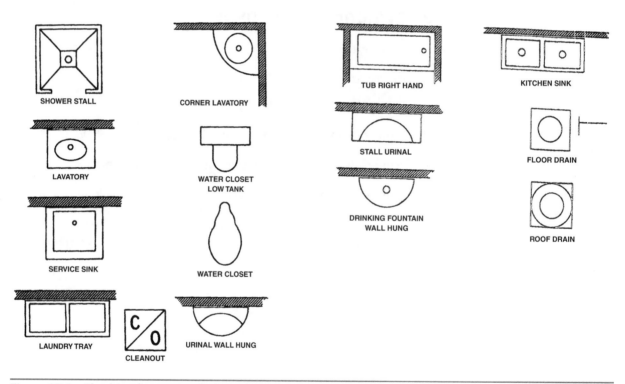

**Figure 17-1**  Plumbing fixtures: symbols.

**Table 17-1** Symbols: Plumbing Piping

| Symbol | Description |
|---|---|
| —·—·—·—·—·— | DOMESTIC COLD WATER, COLD WATER |
| —··—··—··—··—·· | DOMESTIC HOT WATER, HOT WATER |
| —···—···—···—··· | DOMESTIC HOW WATER RETURN, HOT WATER RETURN |
| ----- MA ------ MA ------ MA ----- | MEDICAL AIR |
| —— MV —— MV —— MV —— | MEDICAL VACUUM |
| ----- LA ------ LA ------ LA ----- | LABORATORY AIR |
| —— LV —— LV —— LV —— | LABORATORY VACUUM |
| —— OA —— OA —— OA —— | ORAL EVACUATION |
| ----- IA ------ IA ------ IA ----- | INDUSTRIAL AIR |
| —— D —— D —— D —— | DRAIN |
| -------------------- | VENT (SANITARY) |
| —— SS —— SS —— SS —— | SOIL, WASTE, OR SANITARY SEWER |
| -----SAN------SAN------SAN----- | SANITARY SEWER, BELOW GRADE |
| —— SD —— SD —— SD —— | STORM WATER |
| ----- SD ------ SD ------ SD ----- | STORM WATER, BELOW GRADE |
| ——SCW——SCW——SCW—— | SOFTEN COLD WATER |
| ——FCW——FCW——FCW—— | FILTERED COLD WATER |
| ——DWS——DWS——DWS—— | DRINKING WATER SUPPLY |
| -----DWR------DWR------DWR----- | DRINKING WATER RETURN |
| ——TWS——TWS——TWS—— | TEMPERED WATER SUPPLY |
| -----TWR------TWR------TWR----- | TEMPERED WATER RETURN |
| —— N₂O —— N₂O —— N₂O —— | NITROUS OXIDE |
| —— O —— O —— O —— | OXYGEN |
| —— N₂ —— N₂ —— N₂ —— | NITROGEN |
| —— NG —— NG —— NG —— | NATURAL GAS |
| ----- NG ------ NG ------ NG ----- | NATURAL GAS, BELOW GRADE |
| ——FOD——FOD——FOD—— | FUEL OIL DISCHARGE |
| ——FOS——FOS——FOS—— | FUEL OIL SUPPLY |
| ——FOV——FOV——FOV—— | FUEL OIL VENT |
| -----FOR------FOR------FOR----- | FUEL OIL RETURN |

**Table 17-2**   Symbols: General Plumbing

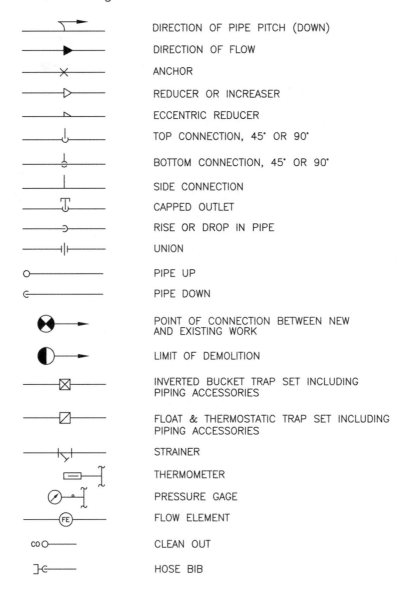

| | |
|---|---|
| | DIRECTION OF PIPE PITCH (DOWN) |
| | DIRECTION OF FLOW |
| | ANCHOR |
| | REDUCER OR INCREASER |
| | ECCENTRIC REDUCER |
| | TOP CONNECTION, 45° OR 90° |
| | BOTTOM CONNECTION, 45° OR 90° |
| | SIDE CONNECTION |
| | CAPPED OUTLET |
| | RISE OR DROP IN PIPE |
| | UNION |
| | PIPE UP |
| | PIPE DOWN |
| | POINT OF CONNECTION BETWEEN NEW AND EXISTING WORK |
| | LIMIT OF DEMOLITION |
| | INVERTED BUCKET TRAP SET INCLUDING PIPING ACCESSORIES |
| | FLOAT & THERMOSTATIC TRAP SET INCLUDING PIPING ACCESSORIES |
| | STRAINER |
| | THERMOMETER |
| | PRESSURE GAGE |
| | FLOW ELEMENT |
| | CLEAN OUT |
| | HOSE BIB |

**Table 17-3**   Symbols: Plumbing Valve

| | |
|---|---|
| ⊳◁ | GATE VALVE |
| ▷◁ | GLOBE VALVE |
| ▷◁—[ | GATE VALVE WITH 3/4 " HOSE ADAPTER |
| ↘ | CHECK VALVE |
| | ANGLE GLOBE VALVE |
| | BUTTERFLY VALVE |
| | BALL VALVE |
| | MODULATING CONTROL VALVE |
| | TWO POSITION CONTROL VALVE |
| | THREE—WAY MODULATING CONTROL VALVE |
| | THREE—WAY TWO POSITION CONTROL VALVE |
| | PRESSURE REGULATING VALVE |
| | AUTOMATIC FLOW CONTROL VALVE |
| | PRESSURE RELIEF VALVE |
| | MANUAL AIR VENT |
| | TEST PLUG (PRESSURE/TEMPERATURE) |
| AV | AUTOMATIC AIR VENT |

## Mathematics

One of the first uses of symbols was in mathematics. Note in Figure 17-2 that the common symbols are pretty much standardized since everyone has been using them all through their education process.

These symbols have been so standardized that they are permanently imprinted on the keys of modern day calculators.

## Abbreviations

To abbreviate means to shorten or make smaller. Table 17-4 lists some of the more common abbreviations used in the building trades and others. Table 17-5 is made up of a group of special abbreviations.

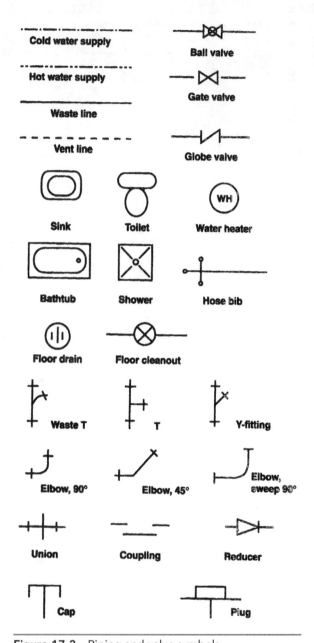

**Figure 17-2**  Piping and valve symbols.

**Table 17-4**  Common Abbreviations

| Plumbing Abbreviations |||||| 
|---|---|---|---|---|---|
| Abbreviation | Meaning | Abbreviation | Meaning | Abbreviation | Meaning |
|---|---|---|---|---|---|
| AD | Area drain | EX | Existing | PPM | Parts per million |
| AFF | Above finish floor | F | Fahrenheit | PSI | Pounds per square inch |
| AFG | Above finish grade | FCO | Floor cleanout | PSIG | Pounds per square in gauge |
| AG | Air gap | FCW | Filtered cold water | PTRV | Pressure temperature relief valve |
| AP | Access panel | FD | Floor drain | PW | Potable water |
| AS | Automatic | FM | Flow meter | RD | Roof drain |
| ASD | Automatic sprinkler drain | FS | Floor sink or flow switch | RDL | Roof drain leader |
| ASHRAE | American Society of Heating, Refrigeration and Air Conditioning | FU | Fixture unit | RL | Roof leader |
| ASME | American Society of Mechanical Engineers | GAL | Gallon | SAN | Sanitary |
| ASPE | American Society of Plumbing Engineers | GPD | Gallons per day | SCW | Softened cold water |
| BFP | Reduced pressure back-flow preventer | GPH | Gallons per hour | SP | Sump pump |
| BT | Bathtub | GPM | Gallons per minute | SPR | Sprinkler line |
| BTU | British thermal unit | GVTR | Gas vent through roof | SQFT | Square feet |
| BTUH | British thermal unit per hour | GWH | Gas-fired water heater | ST | Storage tank |
| C | Celsius | HB | Hose bibb | TCV | Temperature control valve |
| CI | Cast iron | HD | Hub drain | TD | Temperature difference or trench drain |
| CO | Cleanout | HS | Hand sink | TDH | Total dynamic |
| CS | Clinical sink | HST | Hot water storage tank (domestic) | TEMP | Temperature |
| CV | Control valve | HWB | Hot water boiler | TMV | Thermostatic mixing valve |
| DCW | Domestic cold water | HWCP | Hot water circulating pump | TP | Trap primer |
| DHW | Domestic hot water | HWP | Hot water pump | TSTAT | Thermostat |
| DHWR | Domestic hot water return | ICW | Industrial cold water | TWR | Tempered water return |
| DHWS | Domestic hot water supply | IPC | International Plumbing Code | TWS | Tempered water supply |
| DI | Deionized water ACT | IWR | Industrial water return drain | TYP | Typical |
| ESC | Escutcheon | OVFL | Overflow | WH | Water heater or wall hydrant |
| ESH | Emergency shower | PD | Pressure drop or difference | WL | Water line |
| ET | Expansion tank | PDI | Plumbing and Drainage Institute | WM | Water meter |
| EWC | Electric water cooler | PG | Pressure gauge | WPD | Water pressure drop |
| EWH | Electric water heater | PP | Plumbing pump | WS | Waste stack |

Data from the U.S. Department of Veterans Affairs, Plumbing Abbreviations [http://www.cfm.va.gov/til/].

**Table 17-5**    Special Abbreviations

| Abbreviation | Description |
|---|---|
| A or a | Area |
| atm. pres. | Standard atmospheric pressure |
| av. | Average |
| AWG | American wire gauge |
| bbl | Barrels |
| B or b | Breadth |
| bhp | Brake horsepower |
| BM | Board measure |
| Btu | British thermal units |
| BWG | Birmingham wire gauge |
| C of g | Center of gravity |
| cal. val. | Calorific value |
| cm | Centimeters |
| $C_p$ | Specific heat at constant pressure |
| $C_v$ | Specific heat at constant volume |
| cu. | Cubic |
| cu. in. | Cubic inches |
| cyl. | Cylinder |
| D or d | Depth or diameter |
| deg. | Degrees |
| diam. | Diameter |
| evap. | Evaporation |
| F | Fahrenheit |
| g. | Gravity acceleration |
| gals. | Gallons |
| gpm | Gallons per minute |
| H or h | Height, or head of water |
| hp | Horsepower |

| Abbreviation | Description |
|---|---|
| ihp | Indicated horsepower |
| kg | Kilograms |
| lb | Pounds |
| lb-ft | Pound-feet |
| lb per sq. in. | Pounds per square inch |
| log | Logarithm to the base of 10 |
| $log_e$ or ln | Logarithm to the base e |
| min. | Minute |
| mm Hg | Millimeters of mercury (pressure) |
| mol. wt. | Molecular weight |
| OD | Outside diameter (pipes) |
| psi | Pounds per square inch |
| Rad, R, or r | Radius |
| rpm | Revolutions per minute |
| sat | Dry saturated (steam) |
| sec | Second |
| sq. ft or c' | Square feet |
| sq. in. or c" | Square inches |
| sq. yd | Square yards |
| sup | Superheated |
| temp. | Temperature |
| V or v | Velocity |
| V or vol. | Volume |
| WI | Wrought iron |

Data from Babcock & Wilcox Company, Project Gutenberg's Steam, Its Generation and Use (www.gutenberg.org/files/22657/22657-h/chapters/superheat.html) and All Acronyms, Acronyms, Initialisms, Alphabetisms and other Abbreviations (www.all-acronynms.com/).

# Review Questions

1. Where are plumbing symbols used?

   _____

2. Draw a freehand sketch of the symbol that represents a lavatory.

   _____

3. Draw a freehand sketch of the symbol that represents a wall-hung urinal.

   _____

4. Draw a freehand sketch of the symbol of a kitchen sink.

   _____

5. Who standardized the symbols used in plumbing today?

   _____

6. Where were symbols first used?

   _____

7. What is the symbol for square root?

   _____

8. What is the mathematical symbol for a cube?

   _____

9. Write out the equation that shows it represents two numbers that need to be squared.

   _____

10. What does the equation for finding the hypotenuse of a triangle look like?

    _____

# 18 | Metals Encountered in Plumbing

## Performance Objective

After studying this chapter, you will:

- Understand how metal plays an important role in plumbing work.

- Know which metals to avoid in certain situations.

- Know how metals react with one another.

- Be aware of the melting point for metals.

- Be aware of the symbol used to represent specific metals.

- Be able to describe characteristics of lead, potassium and copper.

- Be able to explain how electrolysis take place.

- Be able to answer the review questions at the end of the chapter.

# Metals

Metals cover a large part of the earth. The earth's crust is said to be made up of about 8 percent aluminum, 5 percent iron, and 4 percent calcium. Potassium, sodium, and magnesium also occur in large amounts. The core of the earth is much heavier than the crust. Scientists believe that the core is made up mostly of nickel and iron. They also believe that more heavy metals, such as gold, lead and mercury, are in the core rather than near the surface.

How do you determine what is metallic and what is non-metallic? An electrolysis test is performed to determine which is which. This test dissolves the element in acid and then an electric current is passed through the solution. If the element is metallic, the atoms in it will show a positive charge. This means that when electricity is run through the solution the atoms will seek the negative point where the electricity enters the solution. Chemists define metals as *those elements which, when in solution in a pure state, carry a positive charge and seek the negative pole in an electric cell.* There is one exception, however: hydrogen.

Metals have a silvery color. They are shiny, and usually heavier than water. Most of them conduct heat and electricity. Many of the metals can be hammered into thin sheets. This makes them what is referred to as malleable. They also can be drawn out into wires. This is called a ductile property.

A few metals, such as gold, copper, and strontium, are colored. Some are lighter than water: potassium sodium and lithium will float on water. Some metals are not malleable and ductile, but are brittle. They break easily when worked. Calcium is a good example of a brittle element.

# Non-metals

Boron and selenium are called non-metals. These are not chemically metals, but have one or more of the physical properties of metals.

# Alloys

Combinations of metals are called alloys. Some of these are: bronze, bell metal, gun metal, and type metal. All metals and alloys that do not contain iron are referred to as non-ferrous. Iron has a Latin name of ferrous.

# Substitutes

In many cases metallurgists substitute aluminum for steel. Aluminum is an almost unlimited source of light metal. Magnesium, another light, strong metal is encountered most frequently in aircraft parts. However, it burns quickly and easily. Magnesium is extracted from sea water, but in rock form it is called dolomite. A listing of metals is found in Table 18-1.

# Aluminum

Aluminum is a silvery-white metal. It is light, strong, and rustproof. People often call it the magic metal. It can be stretched and rolled into almost any shape. Aluminum is made into thousands of products. It is used for the bodies and wings of airplanes, thin wrappers for chewing gum and many other products. In the United States, chances are you will very often run into one of the uses for which aluminum is ideally suited. We use more aluminum than any other metal except iron and steel. People of the United States and Canada call this metal aluminum, but people in Great Britain and many other places call it aluminium (pronounced *al-u-min-e-um*).

**Table 18-1**  Metals

| METAL | SYMBOL | SPEC. GRAV. | MELT POINT °C | MELT POINT °F | ELEC. COND. % COPPER | LBS. CU. |
|---|---|---|---|---|---|---|
| ALUMINUM | AL | 2.71 | 660 | 1220 | 64.9 | .0978 |
| ANTIMONY | SB | 6.62 | 630 | 1167 | 4.42 | .2390 |
| ARSENIC | AS | 5.73 | — | — | 4.9 | .2070 |
| BERYLLIUM | BE | 1.83 | 1280 | 2336 | 9.32 | .0660 |
| BISMUTH | BI | 9.80 | 271 | 520 | 1.50 | .3540 |
| BRASS (70–30) | | 8.51 | 900 | 1652 | 28.0 | .3070 |
| BRONZE (5% SN) | | 8.87 | 1000 | 1832 | 18.0 | .3200 |
| CADMIUM | CD | 8.65 | 321 | 610 | 22.7 | .3120 |
| CALCIUM | CA | 1.55 | 850 | 1562 | 50.1 | .0560 |
| COBALT | CO | 8.90 | 1495 | 2723 | 17.8 | .3210 |
| COPPER | CU | | | | | |
| ROLLED | | 8.89 | 1083 | 1981 | 100.0 | .3210 |
| TUBING | | 8.95 | — | — | 100.0 | .3230 |
| GOLD | AU | 19.30 | 1063 | 1945 | 71.2 | .6970 |
| GRAPHITE | | 2.25 | 3500 | 6332 | 10-3 | .0812 |
| INDIUM | IN | 7.30 | 156 | 311 | 20.6 | .2640 |
| IRIDIUM | IA | 22.40 | 2450 | 4442 | 32.5 | .8090 |
| IRON | FE | 7.20 | 1200–1400 | 2192–2552 | 17.6 | .2600 |
| MALLEABLE | | 7.20 | 1500–1600 | 2732–2912 | 10 | .2600 |
| WROUGHT | | 7.70 | 1500–1600 | 2732–2912 | 10 | .2780 |
| LEAD | PB | 11.40 | 327 | 621 | 8.35 | .4120 |
| MAGNESIUM | MG | 1.74 | 651 | 1204 | 38.7 | .0628 |
| MANGANESE | MN | 7.20 | 1245 | 2273 | 0.9 | .2600 |
| MERCURY | HG | 13.65 | -38.9 | -37.7 | 1.80 | .4930 |
| MOLYBDENUM | MO | 10.20 | 2620 | 4748 | 36.1 | .3680 |
| MONEL (63 - 37) | | 8.87 | 1300 | 2372 | 3.0.3200 | |
| NICKEL | NI | 8.90 | 1452 | 2646 | 25.0 | .3210 |
| PHOSPHOROUS | P | 1.82 | 44.1 | 111.4 | 10-17 | .0657 |
| PLATINUM | PT | 21.46 | 1773 | 3221 | 17.5 | .7750 |
| POTASSIUM | K | 0.860 | 62.3 | 144.1 | 28 | .0310 |
| SELENIUM | SE | 4.81 | 220 | 428 | 14.4 | .1740 |
| SILICON | SI | 2.40 | 1420 | 2588 | 10-5 | .0866 |
| SILVER | AG | 10.50 | 960 | 1760 | 106 | .3790 |
| STEEL (CARBON) | | 7.84 | 1330-1380 | 2436-2516 | 10 | .2830 |
| STAINLESS | | | | | | |
| (18-8) | | 7.92 | 1500 | 2732 | 2.5 | .2860 |
| (13-CR) | 7.78 | 1520 | 2768 | 3.5 | .2810 | |
| TANTALUM | TA | 16.60 | 2900 | 5414 | 13.9 | .599 |
| TELLURIUM | TE | 6.20 | 450 | 846 | 10-5 | .224 |
| THORIUM | TH | 11.70 | 1845 | 3353 | 9.10 | .422 |
| TIN | SN | 7.30 | 232 | 449 | 15.00 | .264 |
| TITANIUM | TI | 4.50 | 1800 | 3272 | 2.10 | .162 |
| TUNGSTEN | W | 19.30 | 3410 | — | 31.50 | .697 |
| URANIUM | U | 18.70 | 1130 | 2066 | 2.80 | .675 |
| VANADIUM | V | 5.96 | 1710 | 3110 | 6.63 | .215 |
| ZINC | ZN | 7.14 | 419 | 786 | 29.10 | .258 |
| ZIRCONIUM | ZR | 6.40 | 1700 | 3092 | 4.20 | .231 |

Aluminum has an atomic number of 13, an atomic weight of 26.9815, and the chemical symbol Al. Aluminum weighs only about one-third as much as the same object made of iron or steel. Pure aluminum is soft, but other metals can be alloyed (mixed) with it, to make it as strong as steel.

Alloyed aluminum is used in airplane wings that can withstand loads of more than 90,000 pounds per square inch. Aluminum can be drawn so fine that 1.5 pounds of aluminum wire could circle the earth. Aluminum does not occur as a metal in nature. It must be manufactured from bauxite ore. The United States contributes about 40 percent of the aluminum produced in the world. Russia ranks second. Other leading aluminum producers include Canada, France, Italy, Japan, Norway, and West Germany. Most countries have supplies of bauxite ore. The United States has some bauxite ore, but Canada has none.

## The Importance of Aluminum

Aluminum melts at a higher temperature (about 1120°F) than zinc, tin, and lead, but at a lower temperature than copper and steel. It is an excellent conductor of electricity. Wire used to carry electricity across the country is usually made of aluminum. Exposed aluminum forms a thin coating of oxide that protects the metal against further accumulations and keeps it from rusting.

Aluminum can be produced in more forms than any other metal. Aluminum compounds make up over 15 percent of the earth's crust.

# Copper

The chemical symbol for cooper is Cu. Copper has been used for more than 5,000 years. It is a reddish-orange metal used in a number of applications, in everything from gutters for houses to electronic guidance systems for space rockets.

Copper is the best low-cost conductor of electricity. The electrical industry uses about six-tenths of the copper produced, mainly in the form of copper wire. Copper wire carries most of the electric current for homes, factories, and offices. Large amounts of copper wire are used in telephone and telegraph systems, as well as in television sets, motors, generators, and other kinds of electrical equipment and machinery, not to mention plumbing. Modern house developers use great quantities of copper tubing or pipe for the water systems in the houses they build.

Plastic tubing or piping is replacing the use of copper in most buildings where codes permit.

Combined with other metals, copper forms such alloys as:

- Brass
- Bronze

Chemical compounds are very important because they:

- Improve soil
- Destroy harmful insects

Copper compounds are used in:

- Paint (to protect materials against corrosion)
- Small amounts as it is vital to all plant and animal life

Copper and the symbol Cu come from cuprum, the Roman name for Cyprian metal.

## Properties of Copper

The physical characteristics of copper are:

- Conductivity
- Malleability
- Ductility
- Resistance to corrosion

## Conductivity

Copper is best known for its ability to conduct electricity. Silver is the only better conductor, but silver is too expensive. Copper is also an excellent conductor of heat. Malleability. Pure copper is highly malleable (easy to shape). It does not crack when hammered, stamped, forged, die pressed, or spun into unusual shapes. Copper can be worked (shaped) either hot or cold. It can be rolled into sheets less than 0.05 of an inch thick. Cold rolling changes the physical properties of copper and increases its strength.

**Ductility.** Copper has great ductility, the ability to be drawn into thin wires without breaking.

**Resistance to corrosion.** Copper is quite resistant to corrosion. It will not rust. In damp air, copper turns from a reddish-orange to a reddish-brown color. After long exposure, copper becomes coated with a green film called a patina. The patina protects the copper against further corrosion.

Copper has an atomic number of 29. It has an atomic weight of 63.54. Copper melts at 1083°C and boils at 2595°C. Copper has a density of about 560 pounds per cubic foot and a specific gravity of 8.92.

Most of the copper mined today comes from seven kinds of ores. These ores also may contain other metals, such as lead, zinc, gold, cobalt, platinum, and nickel. Only a small fraction of the copper in use today comes from native or pure copper found in a rare discovery.

# Iron

The chemical symbol for iron is Fe. Iron is a silvery-white metal in its pure state. All plants, animals, and humans need iron in their bodies to help them live. Iron is also important because it is the basic material for many of the manufactured things we use every day. This metal seldom occurs in a pure state. It is obtained from certain kinds of rock, or ore. Iron can take a high polish. It can be welded when hot. Considerable scrap metal is recovered because iron is used widely.

The symbol Fe comes from the Latin word for iron: ferrum. Iron has an atomic weight of and an atomic number of 26. The specific gravity of iron is 7.86 or nearly 8 times as heavy as water. Iron melts at 1535°C and boils at 3000°C. Iron is the main ingredient in steel.

Cast iron refers to a group of ferrous metals with carbon content of 2 to 5 percent. The groups of cast iron are: gray cast iron, white cast iron, malleable iron, ductile (nodular) iron, and alloy cast irons. Gray cast iron is used in bathtubs, cast iron pipe and fittings as well as various other products.

# Steel

Steel is a general term that includes many iron alloys. It is made of iron with a carbon content of 0.05 to 2.0 percent. Carbon steel accounts for 90 percent of the total production. It is used widely in the manufacture of various products. The hardening element for steel is carbon. The amount of carbon in the steel determines its classification as:

- Low carbon steel
- Medium carbon steel
- High carbon steel

Hardness and tensile strength are increased when carbon is added to steel. Ductility and weldability decrease.

# Lead

The symbol for lead is Pb (taken from the Latin name Plumbum). Lead is a soft, heavy metal element. Lead combines easily with other metals to make many of the most useful objects. The Romans used lead for water pipes and other purposes, as did most of Europe until after World War II.

## Properties of Lead

Lead is 11 times as heavy as an equal volume of water. Only gold is heavier than lead. Lead has a bluish-gray color, and is so soft that it can be scratched with your fingernail. When the metal is cut, it gleams silver-white, but it quickly darkens in the air. Lead melts at a temperature of 621°F (327.5°C), which is above the melting point of tin, but below that of zinc. Its atomic weight is 207.19, and its atomic number is 82.

Lead compounds are poisonous. Lead poisoning was once common among persons working with lead. But today, various safeguards have been developed to prevent this. Lead-based paints are no longer permitted. Pewter was made of lead and tin and used as plates for household purposes. This is believed to have caused much of the insanity in early New England.

## Uses of Lead

The pipes that carried water from city mains to household faucets were often made of lead. Lead also was used as a lining for tanks, for insulating sheathing on electric cables, and for plates in automobile storage batteries.

In atomic energy, scientists used lead to shield workers from harmful radiation. Lead alloyed with tin makes solder, a binding agent for uniting several metals, especially copper tubing and fittings. Lead is alloyed with antimony to make bullets, and at one time, printing type. Chemical dyes may be made from lead acetate, or sugar of lead. Carbonate of lead, or white lead, was an ingredient in the paints of yesterday. Red lead, a compound of oxygen and lead, is also used in the making of paints. Lead chromate, usually called chrome-yellow, is another paint pigment. Litharge, another compound of oxygen and lead, is used in making fine glass.

## How Lead Is Obtained

Lead is found in veins and masses in limestone and dolomite. Lead is also found with deposits of other metals such as zinc, silver, copper, and gold. The most common lead-ore mineral is galena, or lead sulfide (chemical symbol, PbS). Another ore mineral in which lead is found combined with sulfur is anglesite, or lead sulfate (PbS04). Cerussite (PbC03) is a mineral that is a carbonate of lead. All three of these ores are found in the United States, which is one of the chief lead-mining countries. The principal lead-mining states include Missouri, Idaho, Utah, and Colorado. Russia and Australia are the leading lead-mining countries. Other important producers include Canada, Mexico, and Peru.

Lead is mined below the surface of the earth. Some veins of lead go as deep as several thousand feet below the surface, but most lead deposits occur close to the surface.

# Magnesium

Magnesium has a grayish-white metallic color. The chemical symbol for magnesium is Mg. Magnesium is found in plentiful amounts in sea water. It is also abundant in the earth's crust, usually combined with other elements. Magnesium is used to protect other metals from rust and corrosion in everything from water heaters to oil tanks. Magnesium burns with a brilliant white light, which makes it useful in flares and tracer bullets. Magnesium is the lightest metal that has been used to build things. It has been used in pure form, but there are many more uses, such as an additive to aluminum steel and other metals.

Magnesium ranks as the third most abundant structural metal in the earth's crust. Only aluminum is more abundant. Magnesium is used to protect pipelines and underground storage tanks. Its atomic number is 12 and its atomic weight is 24.312. Magnesium melts at 651°C. Magnesium has a specific gravity of 1.75.

Magnesium is the lightest of the commercially important metals, having a specific gravity of about 1.75. Two major processes are used to produce magnesium:

- Electrolysis of molten anhydrous magnesium chloride.
- Thermic reduction of dolomite with ferro-silicon.

The magnesium chloride for the first process may be obtained by processing various mineral deposits or by processing seawater, which contains about 1.07 kilograms of magnesium per cubic meter (1 pound per 15 cubic feet).

# Mercury

The chemical symbol for mercury is Hg. Mercury is a silver-white metallic element that flows freely. It was named for the fleet-footed messenger of the gods in Roman mythology. Its fluid quality also gave it the name quicksilver. Mercury was known to ancient Chinese and Hindus and was found in Egyptian tombs from 1500 BC.

Mercury is the only metal that stays liquid at ordinary temperatures. It is 13.55 times as heavy as water. Under ordinary atmospheric conditions, mercury melts at –38.87°C. (about –37.9°F.). Mercury boils at 356.58°C. (about 673°F.). Its atomic weight is 200.59 and its atomic number is 80.

## Expansion and Contraction

Mercury expands and contracts in regular degrees when subjected to changes of temperatures, making it an excellent metal for filling tubes of thermometers and barometers.

Mercury and its compounds are poisonous. However, farmers have previously used mercury compounds to keep fungi from destroying barley and wheat seeds. They also treated apple trees, potato vines, and tomato vines with these compounds. Manufacturers used mercury to keep fungi from growing in such products as paint and paper. Widespread use of mercury and its compounds in agriculture and industry put mercury into the environment. Under certain conditions, this mercury may still contaminate food eaten by animals and human beings.

## Useful Forms

The most useful form of mercury is produced as mercury fulminate. This is a chemical mixture of alcohol, nitric acid, and mercury. It is highly explosive, and is used chiefly for the percussion caps of shells and cartridges. At one time, the most important use of mercury was to extract gold and silver from their ores. The powdered ore was mixed with water and poured over copper plates coated with a thin layer of mercury. The mercury combined with the gold or silver to form amalgam, which remained on the plates while the water washed away the rocks and earth. Then the amalgam was heated until the mercury boiled and became a gas, leaving the gold or silver free.

## Ores

The chief ore mineral of mercury is *cinnabar*, mercuric sulfide. Pure mercury can also be mined. Mercury is removed from its ore by roasting. The cinnabar ore is placed in a stream of air. The heat drives off the sulfur which combines with the oxygen in the air to form a gas.

Spain is the leading producer of mercury. Other countries include China, Italy, and Russia.

In the United States, California, and Nevada also have deposits.

# Oxidation Potential

Reactions of dissimilar metals is of constant concern to plumbers and others who work with metals in any form. Chemists have devised ways to predict the reaction and the degree and speed with which it occurs.

Oxidation potential is a measure, in volts, of an element's tendency to oxidize or lose electrons. The symbol for oxidation potential is E0.

If you know an element's oxidation potential, you can predict how the element will react with other substances. You can also tell how much voltage would be produced if the element were used to make a battery.

Oxidation potentials can be used to predict whether a chemical reaction will take place (Table 18-2).

Chemical reactions take place spontaneously if the total voltage of the reacting element is positive.

Corrosion takes place when one metal loses electrons and another becomes coated with the displaced electrons. A good example is the rusting out of steel on automobiles when salt is used to melt ice on the roads; holes appear where the salt solution (sodium chloride) is sprayed onto the steel with a high ferrous (iron) content. The result is ferric chloride, a very corrosive chemical. Ferric chloride is also used in some places as an etching material of copper to make printed circuit boards for electronic circuits. Proper disposal of these chemicals is of particular concern to plumbers.

**Table 18-2**    Standard Oxidation Potentials

| Metal | E$_0$ (volts) | Loss Rate (lbs/ampere year) |
|---|---|---|
| Copper (Cu) | –0.34 | 45 |
| Iron (Fe) | +0.44 | 20 |
| Magnesium (Mg) | +2.37 | 8.7 |
| Lead (Pb) | +0.13 | 74 |
| Zinc (Zn) | +0.76 | 23.5 |

Data from Clackamas Community College, Standard Oxidation Potentials (http://dl.clackamas.ce.or.us/ch 105-09/standard.htm) and SiliconfarEast.com, Electrode Reduction and Oxidation Potential (http://www.siliconfareast.com/ox_potential.htm).

# Potassium

Potassium has a silver-white metallic color. When a piece of potassium is dropped into water, it quickly releases hydrogen gas and ignites it, causing an explosion. Pure potassium rapidly combines with oxygen and must be stored under a petroleum liquid such as kerosene. This keeps it from combining with moisture or oxygen in the air. Next to lithium, potassium is the lightest metal. Potassium is so soft that it can be cut with a knife.

Potassium is never found as a pure metal. It is always combined with other substances. The pure metal is obtained by passing an electric current through a fused or melted potassium salt compound. The current separates the potassium from the other elements combined with it.

Potassium belongs to the alkali metals group. The chemical symbol for potassium is K. This comes from kalium, the Latin name for the element. Its atomic number is 19 and its atomic weight is 39.0983.

Potassium melts at 63.65°C (146.6°F) and boils at 774°C (1425°F). The metal was first isolated in 1807 by Sir Humphrey Davy, an English chemist.

# Zinc

Zinc is a chemical element. It is a shiny, bluish-white metal. A coating of zinc applied to metals such as iron or steel prevents them from rusting. The coated metal, called galvanized iron or steel, is used in such products as roof gutters and tank linings. Zinc is also used in electric batteries. The chemical symbol for zinc is Zn. Its atomic number is 30 and its atomic weight is 65.37. Zinc melts at 419.4°C. and boils at 907°C. Men have known about zinc for hundreds of years.

Zinc can be combined with other metals to form many alloys (mixtures). For example:

**Table 18-3**  Modern Metals

| Metal | General Description | Use | Use as an Alloy |
|---|---|---|---|
| Aluminum | $\frac{1}{3}$ weight of steel, conductive, corrosion-resistant; 168 lbs; 1220°F | Cooking utensils, electric wire, foil | Magnesium, Manganese, silicon, aircraft alloys |
| Beryllium | High-temperature light metal, scarce, toxic; 112 lbs; 2350°F | Atomic work | High-strength copper and nickel alloys |
| Boron | Hard, crystalline, high-heat and electric resistance; 144 lbs; 4200°F | Oxidation-resistance coating for high-temperature metals | Steel alloys |
| Cadmium | Soft, bright, corrosion-resistant 540 lbs; 1490°F | Plating | Solder alloys atomic-power control |
| Calcium | Soft, white, ductile, light, corrodes rapidly; 97 lbs; 1490°F | Reducing agent in refining of uranium | Reducing agents |
| Cerium | Soft, ductile; 431 lbs; 1184°F | None | Some alloys of steel |
| Cesium | Most reactive of metals; 117 lbs; 83°F | Little | None |
| Chromium | Bright, corrosion-resistant; 443 lbs; 2940°F | Plating | Stainless steels |
| Columbium | Gray, highly stable, scarce; 524 lbs; 3542°F | None | High temperature super alloys |
| Gallium | Low melting point, high boiling point; 369 lbs; 86°F | Special thermometers | Experimental uses |
| Germanium | Grayish crystalline; 334 lbs; 1756°F | Semiconductor diodes | Electronic devices |
| Indium | Soft; 454 lbs; 311°F | Lubricant on bearings | Solders |
| Lithium | Lightest metal, corrodes rapidly; 33 lbs; 367°F | High-temperature greases | Aluminum, zinc, and manganese alloys |
| Magnesium | 1/3 lighter than aluminum; 109 lbs; 1204°F | Chemical | Aluminum, zinc, manganese alloys |
| Molybdenum | Most available high-temperature metal; 636 lbs; 4750°F | High-temperature forgings | Steel alloys |
| Palladium | Close to platinum; 759 lbs; 2831°F | Electrical contacts | Jewelry alloys |
| Plutonium | Radioactive, fissionable secret | Produce in atomic pile | None |
| Polonium | Radioactive; 587 lbs; 1112°F | Atomic uses | None |
| Radium | Radioactive; 312 lbs; 1760°F | Cancer treatment | Luminous paint |
| Rhodium | Hard, unworkable 780 lbs; 3550°F | Electroplating | Platinum alloys |
| Selenium | Brittle, semi-conductive, photosensitive; 299 lbs; 428°F | Electronic rectifiers, photoelectric-eye tubes | Adds machinability to stainless steel and copper |
| Silicon | Most abundant metal, brittle, heat-, corrosion-resistant; 151 lbs; 2590°F | As coating in steel and molybdenum for high-temperature use | As ferrosilicon, adds elasticity to steel |
| Sodium | Soft, highly reactive with water and air; 61 lbs; 208°F | Constituent in making tetraethyl lead | Aluminum-silicon alloys, as liquid-metal heat exchanger in atomic-power plants |
| Strontium | Sister metal to calcium, photoelectric; 158 lbs; 1500°F | As "getter" in vacuum tubes, in blue-green-sensitive photoelectric cells | None |
| Tantalum | Third-highest temperature metal, corrosive-resistant, scarce; 1,036 lbs; 5160°F | In electric grids, anodes, surgical pins, and plates | In chemical heat-transfer equipment |
| Thorium | Radioactive potentially a source of atomic energy; 705 lbs; 3350°F | Pure metal only recently secured | Adds strength to magnesium, life to resistance-heating alloys |

**Table 18-3** Modern Metals (*continued*)

| Metal | General Description | Use | Use as an Alloy |
|-------|---------------------|-----|-----------------|
| Titanium | Lightweight, high strength, corrosive-resistance; 281 lbs; 3300°F | As nonstructural sheet in jet engine shells, ducting, fitting | In high-strength aluminum alloys high-temperature titanium carbides |
| Uranium | Radioactive, fissionable; 1,166 lbs; 2017°F | As source of atomic energy | None |
| Vanadium | Soft, corrosive-resistant; 372 lbs; 3110°F | Pure ductile form only recently achieved, only laboratory uses | Adds extreme toughness to tool-and-die steel, shafts, springs, bearings |
| Zirconium | Sister to titanium high permeability to neutrons; 399 lbs; 3100°F | Ultraductile metal in atomic reactors | Increasingly used as alloy in magnesium, deoxidizes |

- Brass is an alloy of copper and zinc.
- Bronze is copper, tin, and zinc.
- Nickel silver is copper, nickel, and zinc.
- Zinc is also used in solders (easily melted alloys used for joining metals).
- Zinc and its alloys are used in die-casting (forming objects from liquid metal in molds).
- Electroplating is the coating an object by use of electricity.
- Powder metallurgy is the forming of objects from metal powder.

Moist air tarnishes (discolors) zinc with a protective coating of zinc oxide (ZnO). Once a thin layer of this coating forms, air cannot tarnish the zinc below it.

White powdery zinc oxide is one of industry's most useful chemicals. It is used in the manufacture of cosmetics, plastics, rubber, skin ointments, and soaps. It is also used as a pigment (coloring matter) in paints and printing inks. Zinc sulfide (ZnS) glows when ultraviolet light, X-rays, or cathode rays (streams of electrons) shine on it. It is used on luminous dials for clocks. It is also used to coat the inside of television screens and fluorescent lamps. Zinc chloride (ZnCh) in a water solution preserves wood and prevents decay. It also protects the wood from insects.

Zinc metal is never found pure in nature. It occurs combined with sulfur in a mineral called zinc blende or sphalerite. Other zinc-containing minerals include: calamine, franklinite, smithsonite, willemite and zincite. Zinc metal is hard and brittle at room temperature, but it softens when it is heated above 100°C (212°F). It is taken from its ores by heating them in the air to convert them to zinc oxide. The oxide is converted to zinc by heating it with carbon.

## Alkali and Alkaloids

*Alkali* is a term that refers to a group of six chemical elements: lithium sodium, potassium, rubidium, cesium, and francium.

These are called *alkali metals*. Compounds of alkali metals are among the most common and most useful of all chemicals. Industry uses millions of tons of these compounds each year. Sodium hydroxide and potassium hydroxide have many uses in manufacturing processes, especially in their salt form. Glass, paper and textiles, as well as soap, use them in their manufacturing processes. Most of these alkaline compounds are easily dissolved in water and in this form the plumber is usually the person most often in contact with the solutions in their many applications.

**Table 18-4**  Metals with a Future

| Metal | General description | Weight (lbs per cu ft) | Melting point (degrees °F) |
|---|---|---|---|
| Actinium | Radioactive, transitional element | Not established | Not established |
| Americium | Radioactive, artificially made, 1945 | Not established | Not established |
| Barium | Soft, tarnishes in air, toxic | 228 | 1300 |
| Berkelium | Radioactive, artificially made, 1950 | Not established | Not established |
| Californium | Radioactive, artificially made, 1950 | Not established | Not established |
| Curium | Radioactive, artificially made, 1944 | Not established | Not established |
| Dysprosium | Rare earth | 534 | Not established |
| Erbium | Rare earth | 572 | 2282 |
| Europium | Rare earth (used in color TV tubes) | 319 | 2012 |
| Francium | Alkali-type metal, discovered in 1939 | Not established | 73 |
| Gadolinium | Rare earth, highest absorption for slow neutrons | 495 | Not established |
| Hafnium | Close to zirconium, twice as heavy; high electron emission | 830 | 3590 |
| Holmium | Rare earth | Not established | Not established |
| Iridium | Platinum group, hard, brittle | 1399 | 4450 |
| Lanthanum | Among commonest rare earths, ductile | 384 | 1519 |
| Lutecium | Rarest of rare earths | 608 | Not established |
| Neodymium | Rare earth | 434 | 1544 |
| Osmium | Platinum group, hard, unworkable; highest density, small alloy use | 1403 | 4892 |
| Potassium | Alkali metal; soft, reactive | 54 | 1440 |
| Praseodymium | Rare earth | 414 | 1724 |
| Promethium | Rare earth; artificially made | Not established | Not established |
| Protactinium | Radioactive | Not established | 5400 |
| Rhenium | Heavy; used as catalyst | 1281 | 5740 |
| Rubidium | Silvery alkali metal; reactive | 95 | 101 |
| Ruthenium | Platinum group, unworkable | 761 | 4442 |
| Samarium | Rare earth, oxidizes slowly in air | 478 | 2462 |
| Scandium | Rare, only recently isolated | 188 | 2192 |
| Technetium | First artificially made element, 1937 | Not established | 4900 |
| Tellurium | Semimetallic; a semiconductor | 389 | 845 |
| Terbium | Among scarcest rare earths | 529 | 621 |
| Thallium | Soft; between lead and alkali metal | 738 | 578 |
| Thulium | Rare earth | 583 | Not established |
| Ytterbium | Rare earth, close to lutecium | 447 | 3272 |

**Table 18-5   Weights of Steel and Brass Bars**

STEEL—Weights cover hot worked steel about 0.50 percent carbon. One cubic inch weighs 0.2833 lb. High-speed steel, 10 percent heavier.

BRASS—One cubic inch weighs 0.3074 lb.

Actual weight of stock may be expected to vary somewhat from these figures because of variations in manufacturing processes.

| SIZE Inches | Steel lb per ft | | | Brass lb per ft | | | SIZE Inches | Steel lb per ft | | | Brass lb per ft | | |
|---|---|---|---|---|---|---|---|---|---|---|---|---|---|
| | Round | Square | Hex | Round | Square | Hex | | Round | Squre | Hex | Round | Square | Hex |
| 1/16 | 0.0104 | 0.013 | 0.0115 | 0.0113 | 0.0144 | 0.0125 | 13/16 | 1.76 | 2.24 | 1.94 | 1.91 | 2.43 | 2.11 |
| 1/8 | 0.042 | 0.05 | 0.046 | 0.045 | 0.058 | 0.050 | 7/8 | 2.04 | 2.60 | 2.25 | 2.22 | 2.82 | 2.45 |
| 3/16 | 0.09 | 0.12 | 0.10 | 0.102 | 0.130 | 0.112 | 15/16 | 2.35 | 2.99 | 2.59 | 2.55 | 3.24 | 2.81 |
| 1/4 | 0.17 | 0.21 | 0.19 | 0.18 | 0.23 | 0.20 | 1 | 2.67 | 3.40 | 2.94 | 2.90 | 3.69 | 3.19 |
| 5/16 | 0.26 | 0.33 | 0.29 | 0.28 | 0.36 | 0.31 | 1 1/16 | 3.01 | 3.84 | 3.32 | 3.27 | 4.16 | 3.61 |
| 3/8 | 0.38 | 0.48 | 0.42 | 0.41 | 0.52 | 0.45 | 1 1/8 | 3.38 | 4.30 | 3.73 | 3.67 | 4.67 | 4.04 |
| 7/16 | 0.51 | 0.65 | 0.56 | 0.55 | 0.71 | 0.61 | 1 3/16 | 3.77 | 4.80 | 4.16 | 4.08 | 5.20 | 4.51 |
| 1/2 | 0.67 | 0.85 | 0.74 | 0.72 | 0.92 | 0.80 | 1 1/4 | 4.17 | 5.31 | 4.60 | 4.53 | 5.76 | 4.99 |
| 9/16 | 0.85 | 1.08 | 0.94 | 0.92 | 1.17 | 1.01 | 1 5/16 | 4.60 | 5.86 | 5.07 | 4.99 | 6.35 | 5.50 |
| 5/8 | 1.04 | 1.33 | 1.15 | 1.13 | 1.44 | 1.25 | 1 3/8 | 5.04 | 6.43 | 5.56 | 5.48 | 6.97 | 6.04 |
| 11/16 | 1.27 | 1.61 | 1.40 | 1.37 | 1.74 | 1.51 | 1 7/16 | 5.52 | 7.03 | 6.08 | 5.99 | 7.62 | 6.60 |
| 3/4 | 1.50 | 1.92 | 1.66 | 1.63 | 2.07 | 1.80 | 1 1/2 | 6.01 | 7.65 | 6.63 | 6.52 | 8.30 | 7.19 |

Weight of bar 1 foot long

**Table 18-6   Cast Iron Pipe Weight**

| TEN FOOT PIPE BUNDLES NO-HUB PIPE | | | |
|---|---|---|---|
| Size | Pieces | Weight | Height |
| 1½ x 10 | 72 | 2109 | 12" |
| 2 x 10 | 54 | 2068 | 11" |
| 3 x 10 | 36 | 1965 | 14" |
| 4 x 10 | 27 | 1943 | 17" |
| 5 x 10 | 24 | 2380 | 20" |
| 6 x 10 | 18 | 2142 | 23" |
| 8 x 10 | 8 | 1332 | 22" |
| 10 x 10 | 6 | 1542 | 26" |
| 12 x 10 | 6 | 1925 | 29" |
| 15 x 10 | 2 | 1002 | 19" |

Weights are approximate and are for shipping purposes only.
Note: Size is measured in in. x ft.; Weight is measured in lb.
Reproduced with permission of Charlotte Pipe and Foundry Company.

All alkali metals react violently with water, forming hydroxides and releasing hydrogen gas and heat in the process.

*Alkaloid* is a term that refers to a group of organic bases found in plants. They contain carbon, hydrogen, nitrogen and oxygen. Some alkaloids are used for medicines and others are very poisonous to humans and animals. Aconitine, from the aconite plant, is also highly poisonous. Nicotine, another alkaloid, is poisonous and produced by the tobacco plant. It is also used to kill insects.

Morphine and codeine are also known as alkaloids, but have medicinal value. Caffeine from coffee and tea, cocaine from coca, and tubo-curarine from the curare plant are all examples of alkaloids often encountered in the plumbing trade.

# Soldering with Soft Solders

Soft soldering is defined as the bonding of metals together with tin-lead alloys that melt below 800°F. The solder becomes a solvent and is diffused into the metal, making a new intermetallic alloy between the solder and the base metal. Bismuth is also used in some soft solders.

Soft soldering may be performed by using a variety of processes such as:

- Soldering coppers
- Electrically heated soldering irons
- Soldering guns

It may also be performed by:

- Flame heating (natural gas and air-aspirating)

- Dip soldering
- Induction heating
- Resistance heating
- Oven heating
- Ultrasonic soldering

The soldering process is selected by considering the joint design, the product size and the product shape, as well as the number of parts desired.

## Review Questions

1. What metal is heavier than lead?

   _____

2. How does zinc not rust?

   _____

3. What is an alkaloid?

   _____

4. In what metal is a soft solder used?

   _____

5. Why is solder used to keep copper joints together?

   _____

6. In what form is zinc mined?

   _____

7. Where is Bismuth used?

   _____

8. What element is considered a metal, but is not found in a hard form?

   _____

9. What does oxidation potential mean?

   _____

10. What is the symbol for mercury?

   _____

# 19 | Measurement Systems (U.S. to Metric)

## Performance Objectives

After studying this chapter, you will:

- Understand how tables of converted units are produced.

- Know more about using the metric system of measurement.

- Know how to use the charts that have been converted to obtain conversions.

- Know how to convert Fahrenheit to Celsius values.

- Know how to convert pressure from the English unit to the Metric unit.

- Be able to answer the review questions at the end of this chapter.

# Measurement

Measurement makes it possible to construct large buildings and/or small sheds. It is necessary for builders to have the ability to measure and be sure their results are standardized and understood by all in the trade.

Plumbers work with materials that have been standardized in both the U.S. system of measurements with the inch, foot and yard as commonplace, and the metric system with the millimeter, centimeter, and meter as common terms, as well as liter for liquids. Canada and the United Kingdom utilize the metric system, as do all countries on the European continent.

The pascal (Pa) has been adopted for use in checking pressure, however, in the English system it is expressed in psi or pounds per square inch.

# Pressure

The kilopascal (kPa) is the unit of measurement recommended for fluid pressure for almost all fields of use, such as barometric pressure, gas pressure, tire pressure, and water pressure. Table 19-1 shows common pressure conversions. Keep in mind that kilo means 1,000.

When working with pressure, remember that atmospheric pressure is commonly measured at 101.3 kPa metric and 14.7 psi English. A common example of plumbers working with pressure is when they determine pressure for columns of water. When performing pressure calculations for columns of water, keep in mind the following:

■ Atmospheric pressure of 101.3 kPa will balance or support a column of mercury 76 cm high.
■ To find head pressure in decimeters when pressure is given in kPa, divide pressure by 0.9794.
■ To find head pressure in kPa of a column of water given in decimeters, multiply the number of decimeters by 0.0794 (Table 19-1).

# Temperature

Table 19-2 provides conversions of Fahrenheit and Celsius temperatures. There are formulas for making the conversions:

$$F = 1.8 \times C + 32$$

$$c = (F - 32) \div 1.8$$

**Table 19- 1**    Pressure Conversions

| Measurement | Equivalent |
| --- | --- |
| 1 pound per square inch (psi) | 6.894757 kPa |
| 1 m column of water | 9.794 k.Pa (or 0.2476985 kPa per inch) |
| 1 cm column of mercury at 0°C | 1.3332239 kPa |
| 10.2 cm of water | 1 kPa |
| 51 cm of water | 5 kPa |
| 1 inch of mercury (inHg) | 3.386389 kPa |
| 6 cm of mercury | 8 kPa |

Pressure
The kilopascal (kPa) is the unit of measurement recommended for fluid pressure for almost all fields of use, such as barometric pressure, gas pressure, tire pressure, and water pressure.

When working with pressure, remember that atmospheric pressure is commonly measured at 101.3 kPa metric and 14.7 psi English. A common example of plumbers working with pressures is determining pressures for columns of water. When performing pressure calculations for columns of water, keep in mind the following:

• Atmospheric pressure of 101.3 kPa will balance or support a column of mercury 76 cm high.

• To find head pressure in decimeters when pressure is given in kPa, divide pressure by 0.9794.

• To find head pressure in kPa of a column of water given in decimeters, multiply the number of decimeters by 0.9794.

**Table 19- 2**  Temperature Conversions

| CENTRIGRADE AND FAHRENHEIT | | | | | | | |
|---|---|---|---|---|---|---|---|
| DEG-C | DEG-F | DEG·C | DEG-F | DEG-C | DEG-F | DEG·C | DEG-F |
| 0 | 32 | | | | | | |
| 1 | 33.8 | 21 | 69.8 | 41 | 105.8 | 61 | 141.8 |
| 2 | 35.6 | 22 | 71.6 | 42 | 107.6 | 62 | 143.6 |
| 3 | 37.4 | 23 | 73.4 | 43 | 109.4 | 63 | 145.4 |
| 4 | 39.2 | 24 | 75.2 | 44 | 111.2 | 64 | 147.2 |
| 5 | 41 | 25 | 77 | 45 | 113 | 65 | 149 |
| 6 | 42.8 | 26 | 78.8 | 46 | 114.8 | 66 | 150.8 |
| 7 | 44.6 | 27 | 80.6 | 47 | 116.6 67 | 152.6 | |
| 8 | 46.4 | 28 | 82.4 | 48 | 118.4 | 68 | 154.4 |
| 9 | 48.2 | 29 | 84.2 | 49 | 120.2 | 69 | 156.2 |
| 10 | 50 | 30 | 86 | 50 | 122 | 70 | 158 |
| 11 | 51.8 | 31 | 87.8 | 51 | 123.8 | 71 | 159.8 |
| 12 | 53.6 | 32 | 89.6 | 52 | 125.6 | 72 | 161.6 |
| 13 | 55.4 | 33 | 91.4 | 53 | 127.4 | 73 | 163.4 |
| 14 | 57.2 | 34 | 93.2 | 54 | 129.2 | 74 | 165.2 |
| 15 | 59 | 35 | 95 | 55 | 131 | 75 | 167 |
| 16 | 60.8 | 36 | 96.8 | 56 | 132.8 | 76 | 168.8 |
| 17 | 62.6 | 37 | 98.6 | 57 | 134.6 | 77 | 170.6 |
| 18 | 64.4 | 38 | 100.4 | 58 | 136.4 | 78 | 172.4 |
| 19 | 66.2 | 39 | 102.2 | 59 | 138.2 | 79 | 174.2 |
| 20 | 68 | 40 | 104 | 60 | 140 | 80 | 176 |

TEMP. C° = 5/9 x (TEMP. F° – 32)

TEMP. F° = (9/5 x TEMP. C°) + 32

Ambient temperature is the temperature of the surrounding cooling medium.

Rated temperature rise is the permissible rise in temperature above ambient when operating under load.

| Fahrenheit | | Celsius |
|---|---|---|
| 212° | Temperature of boiling water | 100° |
| 176° | | 80° |
| 140° | | 60° |
| 122° | | 50° |
| 104° | | 40° |
| 98.6° | Temperature of human body | 37° |
| 95° | | 35° |
| 86° | | 30° |
| 77° | | 25° |
| 68° | | 20° |
| 50° | | 10° |
| 32° | Temperature of melting ice | 0° |
| –4 | | –20° |
| –40 | Temperature equal | –40° |
| –459.69° | Absolute zero | –273.16° |

These formulas make it easier to convert using a computer or calculator.

## BTUs and Joules

Note how in Table 19-3, a joule is *almost* the same as the BTU, but if you check further you will notice the extra digits past the 1 make a difference. You will see that 144 BTU's and 152 joules are required to change one pound of ice to water—both units of measure are doing the same job. It is usually best to carry the decimal point to five places, especially with calculators and computers that can do it easily.

## The Metric System

One of the most interesting features of the metric system is the fact that it is based on the decimal scale. The metric system uses the decimal system of measures and weights. This system utilizes the meter and the gram as the basic units. The meter is used to measure length. It was intended to be (and is almost) one ten-millionth part of the distance measured on a meridian from the equator to the pole, or 39.37079 inches. Other primary units of measure—the square meter, the cubic meter, the liter, and the gram—are based on the meter.

Illustrated below is the metric system of weights and measures.

- *Milli* indicates the l/l000th part (0.001) of a meter or any other unit being measured, such as a millimeter is 0.001 ampere.
- *Centi* expresses the l/l00th part (0.01) of a meter or any other unit being measured, such as cm or centimeter means one hundredth of a meter.
- *Deci* expresses the 1/10th part of a meter. Not as frequently used as is cm and mm.
- *Deca* expresses 10 times the value. Now, the deca is being used to mean a whole number or 10 times the value stated.
- *Hecto* expresses 100 times the value. It means the *unit* is 100 times the stated value.
- *Kilo* expresses 1,000 times the value.

## Length

Some common metric measurements of length.

- 1 mm = 1 millimeter (mm) = 1 of a meter = 0.03937 in.
- 10 mm = 1 centimeter (cm) = 1/100 of a meter = 0.3937 in.
- 10 cm = 1 decimeter (dm) = UIO of a meter = 3.937 in.
- 10 dm = 1 meter (m) = 1 meter = 39.3707 in. = 3.28 ft.
- 10 m = 1 decameter (da) = 10 meters = 32.809 ft.
- 10 da = 1 hectometer (hm) = 100 meters = 328.09 ft
- 10 hm = l kilometer (km) = 1,000 meters = 0.62137 mile
- 10 km = 1 myriameter = 10,000 meters = 6.2138 mile

## Square Measure

Some common metric measurements of square measure.

- 1 sq. centimeter (cm$^2$) = 0.1550 sq. in.
- 1 sq. decimeter (dm$^2$) = 15.5 sq. in.
- l sq. meter (m$^2$) = 1.196 sq. yd.
- 1 are = 3.954 sq. rd.
- 1 hectare = 2.47 acres
- 1sq. kilometer (km$^2$) = 0.386 sq. mile
- 1 sq. in = 6.452 sq. centimeters (cm$^2$)
- 1sq. ft. = 9.2903 sq. decimeters (dm$^2$)
- 1 sq. yd. = 0.8361 sq. meter (m$^2$)
- 1 sq. rd. = 0.2529 are
- 1 acre = 0.4047 hectare
- 1 sq. mile = 2.59 sq. kilometers (km$^2$)

**Table 19-3**   BTUs and Joules

TWO-WAY CONVERSION TABLE

To convert from the unit of measure in Column B to the unit of measure in Column C, multiply the number of units in Column B by the multiplier in Column A. To convert from Column C to B, use the multiplier in Column D.

EXAMPLE: To convert 1000 BTUs to CALORIES, find the "BTU - CALORIE" combination in Columns B and C. "BTU" is in Column B and "CALORIE" is in Column C; so we are converting from B to C. Therefore, we use Column A multiplier. 1000 BTUs x 251.996 = 251,996 Calories. To convert 251,996 Calories to BTUs, use the same "BTU - CALORIE" combination. But this time you are converting from C to B. Therefore, use Column D multiplier. 251,996 Calories x .0039683 = 1000 BTUs.

| A x B = C | | & | C x D = B |
|---|---|---|---|
| To convert from B to C, Multiply B x A: | | To convert from C to B, Multiply C x D: | |
| A | B | C | D |
| 43,560 | Acre | Sq. Foot | $2.2956 \times 10^{-5}$ |
| $1.5625 \times 10\text{-}3$ | Acre | Sq. Mile | 640 |
| 6.4516 | Ampere per sq. cm. | Ampere per sq. in. | .155003 |
| 1.256637 | Ampere (tum) | Gilberts | 0.79578 |
| 33.89854 | Atmosphere | Foot of H2O | |
| 29.92125 | Atmosphere | Inch of Hg | 0.033421 |
| 14.69595 | Atmosphere | Pound force/sq. in. | 0.06804 |
| 251.996 | BTU | Calorie | $3.96832 \times 10^{-3}$ |
| 778.169 | BTU | Pound-foot force | $1.28507 \times 10^{-3}$ |
| $3.93015 \times 10^{-4}$ | BTU | Horsepower-hour | 2544.43 |
| 1055.056 | BTU | Joule | $9.47817 \times 10^{-4}$ |
| $2.9307 \times 10^{-4}$ | BTU | Kilowatt-hour | 3412.14 |
| $3.93015 \times 10^{-4}$ | BTU/hour | Horsepower | 2544.43 |
| $2.93071 \times 10^{-4}$ | BTU/hour | Kilowatt | 3412.1412 |
| 0.293071 | BTU/hour | Kilowatt | 3.41214 |
| 4.19993 | BTU/minute | Calorie/second | 0.23809 |
| 0.0235809 | BTU/minute | Horsepower | 42.4072 |
| 17.5843 | BTU/minute | Watt | 0.0568 |
| 4.1868 | Calorie | Joule | .238846 |
| 0.0328084 | Centimeter | Foot | 30.48 |
| 0.3937 | Centimeter | Inch | 2.54 |

BTU and Joules

| BTU | Joule |
|---|---|
| 1.0 (mean) | 1.05587 |
| 1.0 (international) | 1.055056 |
| 144 (amount required to change 1 pound of ice to water) | 152 |

## Table Weights

Some common metric measurements of table measure.

- 1 gram = 0.0527 ounce (oz.)
- 1 kilogram (kg) = 2.2046 lbs.
- 1 metric ton = 1.1023 Eng. ton
- 1 ounce (oz.) = 28.35 grams
- 1 lb. = 0.4536 kilogram (kg)
- 1 Eng.ton = 0.9072 metric ton

## Approximate Metric Equivalents

Some common metric equivalents.

- 1 decimeter = 4 inches
- 1 meter = 1.1 yards
- 1 kilometer = $\frac{5}{8}$ mile
- 1 liter = 1 .06 qt. liquid; 0.9 qt dry 1 hectoliter = 2518 bushel
- 1 hectare = $2\frac{1}{2}$ acres
- 1 stere or cu. meter = $\frac{1}{4}$ cord (Note: yes, stere is a measurement—it's one cubic meter)
- 1 kilogram = $2\frac{5}{8}$ lbs.
- 1 metric ton = 2,200 lbs.

## Long Measure

Some common measurements of long distance.

- 12 inches (in. or ") = 1 foot (ft. or ')
- 3 feet ==1 yard (yd.) or 36 inches
- $5\frac{1}{2}$ yards or $16\frac{1}{2}$ feet = 1 rod (rd.)
- 40 rods = 1 furlong (fur) or 660 feet
- 8 furlongs or 320 rods = 1 statute mile (mi) or 5,280 feet

## Nautical Measure

Some common equivalents of nautical measure.

- 6050.26 ft. or 1.15156 statute miles = 1 nautical mile
- 3 nautical miles = 1 league
- 60 nautical miles or 69.1 68 statute miles = 1 degree (at the equator)
- 360 degrees = circumference of earth at the equator

## Square Measure

Some common equivalents of square measure.

- 144 square inches (sq. in.) = 1 square foot (sq. ft.)
- 9 sq. ft. = 1 square yard (sq. yd.)
- $30\frac{1}{4}$ sq. yd. = 1 square rod (sq.rd.) or perch (P)
- 4,840 sq. yds. = 1 acre
- 640 acres = 1 square mile (sq. mi.)

## Calculators

A calculator can be used to improve the accuracy and the speed of calculations. They are inexpensive and easy to operate. The scientific type is best for plumbing work since it has the square root and the reciprocal as well as pi, square, and raising to a power.

All trigonometric functions are also available for easy use with the Hewlett-Packard or Texas Instruments models. Most of the HP and TI models have sufficient functions to solve most plumbing problems.

### Calculator Differences

The Hewlett-Packard varies slightly from the Texas Instrument calculator. The Hewlett-Packard has reverse Polish notation. This means you enter the first number and then follow the formula as written. The Texas Instrument procedure is slightly different. It is suggested that you follow the instruction booklet closely before attempting to perform the calculations.

The calculator you use should have scientific tables, with the ability to do the following:

| Functions | |
|---|---|
| Square Root | [ √ ] |
| Reciprocals | [ 1lx ] |
| Trig Functions | [ cos ] |
| Squares | [ $x^2$ ] |
| Add | [ + ] |
| Subtract | [ − ] |
| Divide | [ ÷ ] |
| Multiply | [ × ] |

The calculator also should have at least one level of memory. It is helpful if the calculator computes pi (π). The solutions to problems can then be more accurate if pi is taken out to six or nine places. On most calculators pi is:

π equals 3.14159217

and is rounded to 3.1416 in most problems.

The number representing pi never comes out even. Some computer experimenters have taken pi out to one million places and it never comes out even.

Other important metric conversion factors are shown in Tables 19-4.

**Table 19-4**   Metric Conversion Factors (Standard Conversion Table— English Tto Metric)

| Symbol | To convert from | Multiply by | To determine | Symbol |
|---|---|---|---|---|
| **LENGTH** | | | | |
| IN | inch | 25.4 | millimeters | mm |
| FT | feet | 0.3048 | meters | m |
| YD | yards | 0.9144 | meters | m |
| MI | miles | 1.609344 | kilometers | km |
| **AREA** | | | | |
| SI | square inches | 645.16 | square millimeters | mm² |
| SF | square feet | 0.09290304 | square meters | m² |
| SY | square yards | 0.83612736 | square meters | mi |
| A | acres | 0.4046856 | hectares | ha |
| MI2 | square miles | 2.59 | square kilometers | km² |
| **VOLUME** | | | | |
| CI | cubic inches | 16.387064 | cubic centimeters | cm³ |
| CF | cubic feet | 0.0283168 | cubic meters | m³ |
| CY | cubic yards | 0.764555 | cubic meters | m³ |
| GAL | gallons | 3.78541 | liters | L |
| OZ | fluid ounces | 0.0295735 | liters | L |
| MBM | thousand feet board | 2.35974 | cubic meters | m3 |
| **MASS** | | | | |
| LB | pounds | 0.4535924 | kilograms | kg |
| TON | short tons (2000 lbs) | 0.9071848 | metric tons | t |
| **PRESSURE AND STRESS** | | | | |
| PSF | pounds per square foot | 47.8803 | pascals | Pa |
| PSI | pounds per square inch | 6.89476 | kilopascals | kPa |
| PSI | pounds per square inch | 0.00689476 | megapascals | Mpa |
| **DISCHARGE** | | | | |
| CFS | cubic feet per second | 0.02831 | cubic meters per second | m³/s |

**Table 19-4**    Metric Conversion Factors (Standard Conversion Table— English Tto Metric) (*continued*)

| Symbol | To convert from | Multiply by | To determine | Symbol |
|---|---|---|---|---|
| | | **VELOCITY** | | |
| FT/SEC | feet per second | 0.3048 | meters per second | m/s |
| | | **INTENSITY** | | |
| IN/HR | inch per hour | 25.4 | millimeters per hour | mm/hr |
| | | **FORCE** | | |
| LB | pound (force) | 4.448222 | newtons | N |
| | | **POWER** | | |
| HP | horsepower | 746.0 | watts | w |
| | | **TEMPERATURE** | | |
| °F | degrees Fahrenheit | 5 x (°F – 32)/9 | degrees Celsius | °C |
| | | **DENSITY** | | |
| lb/ft$^3$ | pounds per cubic foot | 16.01846 | kilograms per cubic meter | kg/m$^3$ |
| | | **ACCELERATION** | | |
| G | freefall, standard | 9.807 | meters per second squared | m/s$^2$ |

To convert from metric to English, divide by the above conversion factors.

# Review Questions

1. Where is the metric system of measure used today?

   _____

2. What are the units of measure for pressure in the metric system?

   _____

3. What is the word stere used for in measuring in the metric system?

   _____

4. What is the value of pi?

   _____

5. What does the unit measure in metric?

   _____

6. How much is something with a preface that reads in deci?

   _____

7. What is the unit pascal used to measure?

   _____

8. How do the Hewlett Packard calculator and the IBM calculator differ?

   _____

# 20 Measuring Weights and Liquids

## *UK and US Systems Plus Metric*

### Performance Objectives

After studying this chapter, you will:

- Understand the difference between the English and metric systems.

- Know how to convert weights and measures from English to metric.

- Know how to convert liquid measurements from English to metric and metric to English.

- Know the difference between red brass and copper pipe.

- Know the welded steel pipe weights and its strength.

- Know the relative weight factor for metal pipe.

- Be able to answer the review questions at the end of this chapter.

# English and UK

One thing to keep in mind: The UK means the United Kingdom, which includes Scotland, Wales, and England itself. Sometimes the unit of measure is referred to as the English method. In most cases the English and UK are interchangeable. Many of the weights and measures utilized years ago have still hung around through history and you will find them in other systems of measure.

The liter (litre in Europe) as defined in 1964 by the Conference Genrale des poids et Measures, is equal to one cubic decimeter. The milliliter is equal to one cubic centimeter.

The peck and bushel can no longer be used for trade purposes in the United Kingdom.

The minim and fluid drachm can no longer be used for trade purposes in the United Kingdom.

The UK dry pint and dry quart have the same capacity as their counterparts in liquid measure, but US dry pints and quarts are not equivalents of US liquid pints and quarts. The UK barrel (used in liquid measures only) varies in capacity between 31.5 and 36 UK gallons, depending on what is being measured. As a unit of measurement for fruit and vegetables and dried commodities, the UK dry barrel is equal to 105 US dry quarts.

- The grain has the same value in avoirdupois, troy and apothecaries' systems.
- The ounce troy and apothecaries' ounce are identical and differ from the avoirdupois ounce.
- The troy pound has no legal basis in the United Kingdom but is legalized in the United States.
- The apothecaries' units are now illegal for use in the United Kingdom.

As you can see, measurement units have evolved over time and are still in the process of evolving. A good example of this is the fact that Canada has converted to the metric system, but the United States has stuck with the "English" system of inches, feet, yards, pounds, and ounces (the same as before the 1970s push for conversion to metric).

The Tables 20-1 to 20-6 indicate the various relationships between what you would find in the United Kingdom and the United States. Look them over and study them, for they will be very important in your career as you begin to navigate the world of measurement.

**Table 20-1**   Metric System

| PREFIXES | | | | | |
|---|---|---|---|---|---|
| A. MEGA | = | 1,000,000 | E. DECI | = | 0.1 |
| B. KILO | = | 1000 | F. CENTI | = | 0.01 |
| C. HECTO | = | 100 | G. MILLI | = | 0.001 |
| D. DEKA | = | 10 | H. MICRO | = | 0.000001 |
| **LINEAR MEASURE** | | | | | |
| (The unit is the meter = 39.37 inches) | | | | | |
| 1 CENTIMETER | = | 10 MILLIMETERS | = | 0.3937011 IN. | |
| 1 DECIMETER | = | 10 CENTIMETERS | = | 3.9370113 INS. | |
| 1 METER | = | 10 DECIMETERS | = | 1.0936143 YDS. | |
| 1 DEKAMETER | = | 10 METERS | = | 10.936143 YDS. | |
| 1 HECTOMETER | = | 10 DEKAMETERS | = | 109.36143 YDS. | |
| 1 KILOMETER | = | 10 HECTOMETERS | = | 0.62137 MILE | |
| 1 MYRIAMETER | = | 10,000 METERS | | | |
| **SQUARE MEASURE** | | | | | |
| (The unit is the square meter = 1549.9969 sq. inches) | | | | | |
| 1 SQ. CENTIMETER | = | 100 SQ. MILLIMETERS | = | 0.1550 SQ. IN. | |
| 1 SQ. DECIMETER | = | 100 SQ. CENTIMETERS | = | 15.550 SQ. INS. | |
| 1 SQ. METER | = | 100 SQ. DECIMETERS | = | 10.7639 SQ. FT. | |
| 1 SQ. DEKAMETER | = | 100 SQ. METERS | = | 119.60 SQ. YDS. | |
| 1 SQ. HECTOMETER | = | 100 SQ. DEKAMETERS | | | |
| 1 SQ. KILOMETER | = | 100 SQ. HECTOMETERS | | | |
| (The unit is the "are" = 100 sq. meters) | | | | | |
| 1 CENTIARE | = | 10 MILLIARES | = | 10.7643 SQ. FT. | |
| 1 DECIARE | = | 10 CENTIARES | = | 11.96033 SQ. YDS. | |
| 1 ARE | = | 10 DECIARES | = | 119.6033 SQ. YDS. | |
| 1 DEKARE | = | 10 ARES | = | 0.247110 ACRES | |
| 1 HEKTARE | = | 10 DEKARES | = | 2.471098 ACRES | |
| 1 SQ. KILOMETER | = | 100 HECTARES | = | 0.38611 SQ. MILE | |
| **CUBIC MEASURE** | | | | | |
| 1 DECISTERE | = | 10 CENTISTERES | = | 3.531562 CU. FT. | |
| 1 STERE | = | 10 DECISTERES | = | 1.307986 CU. YDS. | |
| 1 DEKASTERE | = | 10 STERES | = | 13.07986 CU. YDS. | |

**Table 20-2**   Weights and Measures

| LINEAR MEASURE | | | | |
|---|---|---|---|---|
| | | 1 INCH | = | 2.540 CENTIMETERS |
| 12 INCHES | = | 1 FOOT | = | 3.048 DECIMETERS |
| 3 FEET | = | 1 YARD | = | 9.144 DECIMETERS |
| 5.5 YARDS | = | 1 ROD | = | 5.029 METERS |
| 40 RODS | = | 1 FURLONG | = | 2.018 HECTOMETERS |
| 8 FURLONGS | = | 1 MILE | = | 1.609 KILOMETERS |

| MILE MEASUREMENTS | | |
|---|---|---|
| 1 STATUTE MILE | = | 5,280 FEET |
| 1 SCOTS MILE | = | 5,952 FEET |
| 1 IRISH MILE | = | 6,720 FEET |
| 1 RUSSIAN VERST | = | 3,504 FEET |
| 1 ITALIAN MILE | = | 4,401 FEET |
| 1 SPANISH MILE | = | 15,084 FEET |

| OTHER LINEAR MEASUREMENTS | | | | | |
|---|---|---|---|---|---|
| 1 HAND | = | 4 INCHES | 1 FATHOM | = | 6 FEET |
| 1 SPAN | = | 9 INCHES | 1 FURLONG | = | 10 CHAINS |
| 1 CHAIN | = | 22 YARDS | 1 CABLE | = | 608 FEET |
| 1 LINK | = | 7.92 INCHES | | | |

| SQUARE MEASURE | | |
|---|---|---|
| 144 SQUARE INCHES | = | 1 SQUARE FOOT |
| 9 SQUARE FEET | = | 1 SQUARE YARD |
| 30¼ SQUARE YARDS | = | 1 SQUARE ROD |
| 40 RODS | = | 1 ROOD |
| 4 ROODS | = | 1 ACRE |
| 640 ACRES | = | 1 SQUARE MILE |
| 1 SQUARE MILE | = | 1 SECTION |
| 36 SECTIONS | = | 1 TOWNSHIP |

| CUBIC OR SOLID MEASURE | | |
|---|---|---|
| 1 CU.FOOT | = | 1728 CU. INCHES |
| 1 CU. YARD | = | 27 CU. FEET |
| 1 CU. FOOT | = | 7.48 GALLONS |
| 1 GALLON (WATER) | = | 8.34 LBS. |
| 1 GALLON (U.S.) | = | 231 CU. INCHES OF WATER |
| 1 GALLON (IMPERIAL) | = | 277¼ CU. INCHES OF WATER |

| LIQUID MEASUREMENTS | | |
|---|---|---|
| 1 PINT | = | 4 GILLS |
| 1 QUART | = | 2 PINTS |
| 1 GALLON | = | 4 QUARTS |
| 1 FIRKIN | = | 9 GALLONS (ALE OR BEER) |
| 1 BARREL | = | 42 GALLONS (PETROLEUM OR CRUDE OIL) |

**Table 20-2**  Weights and Measures (*continued*)

| DRY MEASURE | | |
|---|---|---|
| 1 QUART | = | 2 PINTS |
| 1 PECK | = | 8 QUARTS |
| 1 BUSHEL | = | 4 PECKS |
| **WEIGHT MEASUREMENT (MASS)** | | |
| **A. AVOIRDUPOIS WEIGHT:** | | |
| 1 OUNCE | = | 16 DRAMS |
| 1 POUND | = | 16 OUNCES |
| 1 HUNDREDWEIGHT | = | 100 POUNDS |
| 1 TON | = | 2000 POUNDS |
| **B. TROY WEIGHT:** | | |
| 1 CARAT | = | 3.17 GRAINS |
| 1 PENNYWEIGHT | = | 20 GRAINS |
| 1 OUNCE | = | 20 PENNYWEIGHTS |
| 1 POUND | = | 12 OUNCES |
| 1 LONG HUNDRED-WEIGHT | = | 112 POUNDS |
| 1 LONG TON | = | 20 LONG HUNDREDWEIGHTS |
|  | = | 2240 POUNDS |

| **C. APOTHECARIES WEIGHT:** | | | | |
|---|---|---|---|---|
| 1 SCRUPLE | = | 20 GRAINS | = | 1.296 GRAMS |
| 1 DRAM | = | 3 SCRUPLES | = | 3.888 GRAMS |
| 1 OUNCE | = | 3 DRAMS | = | 31.1035 GRAMS |
| 1 POUND | = | 12 OUNCES | = | 373.2420 GRAMS |

| **D. KITCHEN WEIGHTS AND MEASURES:** | | |
|---|---|---|
| 1 U.S. PINT | = | 16 FL. OUNCES |
| 1 STANDARD CUP | = | 8 FL. OUNCES |
| 1 TABLESPOON | = | 05. FL. OUNCES (15 CU. CMS.) |
| 1 TEASPOON | = | 0.16 FL. OUNCES (5 CU. CMS.) |

**Table 20-3**   Brass Pipe

| WEIGHT OF SEAMLESS BRASS AND COPPER PIPE<br>Iron Pipe Sizes<br>REGULAR PIPE | | | | | |
|---|---|---|---|---|---|
| **Pipe Size (Inches)** | **DIMENSIONS IN INCHES** | | | **POUNDS PER LINEAL FOOT** | |
| | **O.D.** | **I.D.** | **Wall Thickness** | **Red Brass** | **Copper** |
| 1/8 | .405 | .281 | .062 | .253 | .259 |
| 1/4 | .540 | .376 | .082 | .447 | .457 |
| 3/8 | .675 | .495 | .090 | .627 | .641 |
| 1/2 | .840 | .626 | .107 | .934 | .955 |
| 3/4 | 1.050 | .822 | .114 | 1.27 | 1.30 |
| 1 | 1.315 | 1.063 | .126 | 1.78 | 1.82 |
| 1 1/4 | 1.660 | 1.368 | .146 | 2.63 | 2.69 |
| 1 1/2 | 1.900 | 1.600 | .150 | 3.13 | 3.20 |
| 2 | 2.375 | 2.063 | .156 | 4.12 | 4.22 |
| 2 1/2 | 2.875 | 2.501 | .187 | 5.99 | 6.12 |
| 3 | 3.500 | 3.062 | .219 | 8.56 | 8.75 |
| 3 1/2 | 4.000 | 3.500 | .250 | 11.2 | 11.4 |
| 4 | 4.500 | 4.000 | .250 | 12.7 | 12.9 |
| 5 | 5.563 | 5.063 | .250 | 15.8 | 16.2 |
| 6 | 6.625 | 6.125 | .250 | 19.0 | 19.4 |
| WEIGHT OF SEAMLESS BRASS AND COPPER PIPE<br>EXTRA STRONG PIPE | | | | | |
| **Pipe Size (Inches)** | **DIMENSIONS IN INCHES** | | | **POUNDS PER LINEAL FOOT** | |
| | **O.D.** | **I.D.** | **Wall Thickness** | **Red Brass** | **Copper** |
| 1/8 | .405 | .205 | .100 | .363 | .371 |
| 1/4 | .540 | .294 | .123 | .611 | .625 |
| 3/8 | .675 | .421 | .127 | .829 | .847 |
| 1/2 | .840 | .542 | .149 | 1.23 | 1.25 |
| 3/4 | 1.050 | .736 | .157 | 1.67 | 1.71 |
| 1 | 1.315 | .951 | .182 | 2.46 | 2.51 |
| 1 1/4 | 1.660 | 1.272 | .194 | 3.39 | 3.46 |
| 1 1/2 | 1.900 | 1.494 | .203 | 4.10 | 4.19 |
| 2 | 2.375 | 1.933 | .221 | 5.67 | 5.80 |
| 2 1/2 | 2.875 | 2.315 | .280 | 8.66 | 8.85 |
| 3 | 3.500 | 2.892 | .304 | 11.6 | 11.8 |

**Table 20-3**   Brass Pipe (*continued*)

| RED BRASS 85% Pipe with Threaded Ends Allowable Pressure | | | |
|---|---|---|---|
| **POUNDS PER SQUARE INCH** | | | |
| **Standard Size (Inches)** | **@ 100°F** | **@ 200°F** | **@ 300°F** | **@ 400°F** |
| 1/8 | 370 | 370 | 320 | 140 |
| 1/4 | 870 | 870 | 760 | 330 |
| 3/8 | 890 | 890 | 780 | 340 |
| 1/2 | 900 | 900 | 790 | 340 |
| 3/4 | 810 | 810 | 710 | 310 |
| 1 | 630 | 630 | 560 | 240 |
| 1 1/4 | 690 | 690 | 610 | 260 |
| 1 1/2 | 630 | 630 | 560 | 240 |
| 2 | 540 | 540 | 480 | 210 |
| 2 1/2 | 450 | 450 | 390 | 170 |
| 3 | 510 | 510 | 450 | 190 |
| 3 1/2 | 570 | 570 | 500 | 220 |
| 4 | 510 | 510 | 440 | 190 |
| 5 | 410 | 410 | 360 | 160 |
| 6 | 340 | 340 | 300 | 130 |

| RED BRASS 85% Pipe with Threaded Ends EXTRA STRONG (XS) Allowable Internal Pressure | | | |
|---|---|---|---|
| **POUNDS PER SQUARE INCH** | | | |
| **Standard Size (Inches)** | **@ 100°F** | **@ 200°F** | **@ 300°F** | **@ 400°F** |
| 1/8 | 1960 | 1960 | 1710 | 740 |
| 1/4 | 2210 | 2210 | 1930 | 830 |
| 3/8 | 1840 | 1840 | 1610 | 690 |
| 1/2 | 1760 | 1760 | 1540 | 660 |
| 3/4 | 1510 | 1510 | 1320 | 570 |
| 1 | 1340 | 1340 | 1180 | 510 |
| 1 1/4 | 1160 | 1160 | 1020 | 440 |
| 1 1/2 | 1090 | 1090 | 960 | 410 |
| 2 | 1000 | 1000 | 870 | 380 |
| 2 1/2 | 970 | 970 | 850 | 370 |
| 3 | 910 | 910 | 790 | 340 |

**Table 20-4** Carbon Steel Grade B Pipes

CARBON STEEL GRADE B PIPES - ASTM A53M, A106M, API 5L, Seamless

| Nominal Size | | Outside Diameter (mm) | Schedule | | Wall Thickness (mm) | Maximum Allowable Operating Pressure (MPa) Temperature (°C) | | | | | | |
|---|---|---|---|---|---|---|---|---|---|---|---|---|
| (DN) | (NPS) | | | | | −29 to +38 | 204 | 260 | 343 | 371 | 399 | 427 |
| | | | | | | Maximum Allowable Stress (MPa) | | | | | | |
| | | | | | | 137.9 | 137.9 | 130.3 | 117.2 | 113.8 | 89.6 | 74.5 |
| 15 | ½ | 21.3 | STD | 40 | 2.77 | 34.5 | 34.5 | 32.6 | 29.3 | 28.5 | 22.3 | 18.6 |
| 20 | ¾ | 26.7 | STD | 40 | 2.87 | 28.1 | 28.1 | 26.5 | 23.8 | 23.1 | 18.2 | 15.1 |
| | | | XS | 80 | 3.91 | 39.4 | 39.4 | 37.2 | 33.5 | 32.5 | 25.6 | 21.3 |
| 25 | 1 | 33.4 | STD | 40 | 3.38 | 26.3 | 26.3 | 24.8 | 22.3 | 21.7 | 17.1 | 14.2 |
| | | | XS | 80 | 4.55 | 26.3 | 36.3 | 34.3 | 30.9 | 30.0 | 23.6 | 19.6 |
| 32 | 1¼ | 42.2 | STD | 40 | 3.56 | 21.6 | 21.6 | 20.4 | 18.4 | 17.8 | 14.1 | 11.7 |
| | | | XS | 80 | 4.85 | 30.2 | 30.2 | 28.5 | 25.6 | 24.9 | 19.6 | 16.3 |
| | | | | 160 | 6.35 | 40.6 | 40.6 | 38.4 | 34.5 | 33.5 | 26.4 | 21.9 |
| 40 | 1½ | 48.3 | STD | 40 | 3.68 | 19.4 | 19.4 | 18.4 | 16.5 | 16.0 | 12.6 | 10.5 |
| | | | XS | 80 | 5.08 | 27.4 | 27.4 | 25.9 | 23.3 | 22.6 | 17.8 | 14.8 |
| | | | | 160 | 7.14 | 39.8 | 39.8 | 37.6 | 33.8 | 32.8 | 25.9 | 21.5 |
| 50 | 2 | 60.3 | STD | 40 | 3.91 | 16.4 | 16.4 | 15.5 | 13.9 | 13.5 | 10.7 | 8.9 |
| | | | XS | 80 | 5.54 | 23.7 | 23.7 | 22.4 | 20.1 | 19.5 | 15.4 | 12.8 |
| | | | | 160 | 8.74 | 38.9 | 38.9 | 36.8 | 33.1 | 32.1 | 25.3 | 21.0 |
| 65 | 2½ | 73.0 | STD | 40 | 5.16 | 17.9 | 17.9 | 17.0 | 15.3 | 14.8 | 11.7 | 9.7 |
| | | | XS | 80 | 7.01 | 24.8 | 24.8 | 23.5 | 21.1 | 20.5 | 16.1 | 13.4 |
| | | | | 160 | 9.53 | 34.7 | 34.7 | 32.8 | 29.5 | 28.6 | 22.5 | 18.7 |
| 80 | 3 | 88.9 | STD | 40 | 5.49 | 15.6 | 15.6 | 14.7 | 13.2 | 12.8 | 10.1 | 8.4 |
| | | | XS | 80 | 7.62 | 22.0 | 22.0 | 20.8 | 18.7 | 18.2 | 14.3 | 11.9 |
| | | | | 160 | 11.13 | 33.1 | 33.1 | 31.3 | 28.1 | 27.3 | 21.5 | 17.9 |
| 100 | 4 | 114.3 | STD | 40 | 6.02 | 13.2 | 13.2 | 12.5 | 11.2 | 10.9 | 8.6 | 7.1 |
| | | | XS | 80 | 8.56 | 19.1 | 19.1 | 18.0 | 16.2 | 15.7 | 12.4 | 10.3 |
| | | | | 120 | 11.13 | 25.2 | 25.2 | 23.8 | 21.4 | 20.8 | 16.4 | 13.6 |
| | | | | 160 | 13.49 | 31.0 | 31.0 | 29.3 | 26.4 | 25.6 | 20.2 | 16.8 |
| | | | XXS | | 17.12 | 40.4 | 40.4 | 38.2 | 34.3 | 33.3 | 26.2 | 21.8 |
| 125 | 5 | 141.3 | STD | 40 | 6.55 | 11.6 | 11.6 | 10.9 | 9.8 | 9.5 | 7.5 | 6.2 |
| | | | XS | 80 | 9.53 | 17.1 | 17.1 | 16.1 | 14.5 | 14.1 | 11.1 | 9.2 |
| | | | | 120 | 12.7 | 23.1 | 23.1 | 21.9 | 19.7 | 19.1 | 15.0 | 12.5 |
| | | | | 160 | 15.88 | 29.4 | 29.4 | 27.8 | 25.0 | 24.3 | 19.1 | 15.9 |
| | | | XXS | | 10.05 | 35.9 | 35.9 | 33.9 | 30.5 | 29.6 | 23.4 | 19.4 |
| 150 | 6 | 168.3 | STD | 40 | 7.11 | 10.5 | 10.5 | 9.9 | 8.9 | 8.7 | 6.8 | 5.7 |
| | | | XS | 80 | 10.97 | 16.5 | 16.5 | 15.6 | 14.0 | 13.6 | 10.7 | 8.9 |
| | | | | 120 | 14.27 | 21.8 | 21.8 | 20.6 | 18.5 | 17.9 | 14.1 | 11.7 |
| | | | XXS | 160 | 18.26 | 28.3 | 28.3 | 26.8 | 24.1 | 23.4 | 18.4 | 15.3 |

**Table 20-4** Carbon Steel Grade B Pipes (*continued*)

| Nominal Size | | Outside Diameter (mm) | Schedule | | Wall Thickness (mm) | Temperature (°C) | | | | | | |
|---|---|---|---|---|---|---|---|---|---|---|---|---|
| | | | | | | −29 to +38 | 204 | 260 | 343 | 371 | 399 | 427 |
| | | | | | | Maximum Allowable Stress (MPa) | | | | | | |
| (DN) | (NPS) | | | | | 137.9 | 137.9 | 130.3 | 117.2 | 113.8 | 89.6 | 74.5 |
| 200 | 8 | 219.1 | | 20 | 6.35 | 7.1 | 7.1 | 6.7 | 6.1 | 5.9 | 4.6 | 3.9 |
| | | | | 30 | 7.04 | 7.9 | 7.9 | 7.5 | 6.7 | 6.5 | 5.2 | 4.3 |
| | | | STD | 40 | 8.18 | 9.3 | 9.3 | 8.7 | 7.9 | 7.6 | 6.0 | 5.0 |
| | | | | 60 | 10.31 | 11.7 | 11.7 | 11.1 | 10.0 | 9.7 | 7.6 | 6.3 |
| | | | XS | 80 | 12.7 | 14.6 | 14.6 | 13.8 | 12.4 | 12.0 | 9.5 | 7.9 |
| | | | | 100 | 15.09 | 17.5 | 17.5 | 16.5 | 14.8 | 14.4 | 11.4 | 9.4 |
| | | | | 120 | 18.26 | 21.4 | 21.4 | 20.2 | 18.2 | 17.6 | 13.9 | 11.5 |
| | | | | 140 | 20.62 | 24.3 | 24.3 | 23.0 | 20.7 | 20.1 | 15.8 | 13.1 |
| | | | XXS | | 22.23 | 26.4 | 26.4 | 24.9 | 22.4 | 21.7 | 17.1 | 14.2 |
| | | | | 160 | 23.01 | 27.4 | 27.4 | 25.8 | 23.3 | 22.6 | 17.8 | 14.8 |
| 250 | 10 | 273.1 | | 20 | 6.35 | 5.7 | 5.7 | 5.4 | 4.8 | 4.7 | 3.7 | 3.1 |
| | | | | 30 | 7.8 | 7.0 | 7.0 | 6.6 | 6.0 | 5.8 | 4.6 | 3.8 |
| | | | STD | 40 | 9.27 | 8.4 | 8.4 | 7.9 | 7.1 | 6.9 | 5.5 | 4.5 |
| | | | XS | 60 | 12.7 | 11.6 | 11.6 | 11.0 | 9.9 | 9.6 | 7.5 | 6.3 |
| | | | | 80 | 15.09 | 13.9 | 13.9 | 13.1 | 11.8 | 11.4 | 9.0 | 7.5 |
| | | | | 100 | 18.26 | 16.9 | 16.9 | 16.0 | 14.4 | 14.0 | 11.0 | 9.1 |
| | | | | 120 | 21.44 | 20.0 | 20.0 | 18.9 | 17.0 | 16.5 | 13.0 | 10.8 |
| | | | XXS | 140 | 25.4 | 24.0 | 24.0 | 22.7 | 20.4 | 19.8 | 15.6 | 13.0 |
| | | | | 160 | 28.58 | 27.3 | 27.3 | 25.8 | 23.2 | 22.5 | 17.7 | 14.7 |
| 300 | 12 | 323.9 | | 20 | 6.35 | 4.8 | 4.8 | 4.5 | 4.1 | 4.0 | 3.1 | 2.6 |
| | | | | 30 | 8.38 | 6.4 | 6.4 | 6.0 | 5.4 | 5.2 | 4.1 | 3.4 |
| | | | STD | | 9.53 | 7.2 | 7.2 | 6.9 | 6.2 | 6.0 | 4.7 | 3.9 |
| | | | | 40 | 10.31 | 7.9 | 7.9 | 7.4 | 6.7 | 6.5 | 5.1 | 4.2 |
| | | | XS | | 12.7 | 9.7 | 9.7 | 9.2 | 8.3 | 8.0 | 6.3 | 5.3 |
| | | | | 60 | 14.27 | 11.0 | 11.0 | 10.4 | 9.3 | 9.0 | 7.1 | 5.9 |
| | | | | 80 | 17.48 | 13.5 | 13.5 | 12.8 | 11.5 | 11.2 | 8.8 | 7.3 |
| | | | | 100 | 21.44 | 16.8 | 16.8 | 15.8 | 14.2 | 13.8 | 10.9 | 9.0 |
| | | | XXS | 120 | 25.4 | 20.0 | 20.0 | 18.9 | 17.0 | 16.5 | 13.0 | 10.8 |
| | | | | 140 | 28.58 | 22.7 | 22.7 | 21.4 | 19.3 | 18.7 | 14.8 | 12.3 |
| | | | | 160 | 33.32 | 26.8 | 26.8 | 25.3 | 22.7 | 22.1 | 17.4 | 14.4 |

**Table 20-4** Carbon Steel Grade B Pipes (*continued*)

| Nominal Size (DN) | (NPS) | Outside Diameter (mm) | Schedule | | Wall Thickness (mm) | −29 to +38 / 137.9 | 204 / 137.9 | 260 / 130.3 | 343 / 117.2 | 371 / 113.8 | 399 / 89.6 | 427 / 74.5 |
|---|---|---|---|---|---|---|---|---|---|---|---|---|
| 350 | 14 | 355.6 | | 10 | 6.35 | 4.4 | 4.4 | 4.1 | 3.7 | 3.6 | 2.8 | 2.4 |
| | | | | 20 | 7.92 | 5.5 | 5.5 | 5.2 | 4.6 | 4.5 | 3.5 | 2.9 |
| | | | STD | 30 | 9.53 | 6.6 | 6.6 | 6.2 | 5.6 | 5.4 | 4.3 | 3.6 |
| | | | | 40 | 11.13 | 7.7 | 7.7 | 7.3 | 6.6 | 6.4 | 5.0 | 4.2 |
| | | | XS | | 12.7 | 8.8 | 8.8 | 8.4 | 7.5 | 7.3 | 5.7 | 4.8 |
| | | | | 60 | 15.09 | 10.6 | 10.6 | 10.0 | 9.0 | 8.7 | 6.9 | 5.7 |
| | | | | 80 | 19.05 | 13.4 | 13.4 | 12.7 | 11.4 | 11.1 | 8.7 | 7.3 |
| | | | | 100 | 23.83 | 17.0 | 17.0 | 16.0 | 14.4 | 14.0 | 11.0 | 9.2 |
| | | | | 120 | 27.79 | 20.0 | 20.0 | 18.9 | 17.0 | 16.5 | 13.0 | 10.8 |
| | | | | 140 | 31.75 | 23.0 | 23.0 | 21.7 | 19.5 | 19.0 | 14.9 | 12.4 |
| | | | | 160 | 35.71 | 26.1 | 26.1 | 24.6 | 22.2 | 21.5 | 16.9 | 14.1 |
| 400 | 16 | 406.4 | | 10 | 6.35 | 3.8 | 3.8 | 3.6 | 3.2 | 3.1 | 2.5 | 2.1 |
| | | | | 20 | 7.92 | 4.8 | 4.8 | 4.5 | 4.1 | 3.9 | 3.1 | 2.6 |
| | | | STD | 30 | 9.53 | 5.8 | 5.8 | 5.4 | 4.9 | 4.7 | 3.7 | 3.1 |
| | | | XS | 40 | 12.7 | 7.7 | 7.7 | 7.3 | 6.6 | 6.4 | 5.0 | 4.2 |
| | | | | 60 | 16.66 | 10.2 | 10.2 | 9.6 | 8.7 | 8.4 | 6.6 | 5.5 |
| | | | | 80 | 21.44 | 13.2 | 13.2 | 12.5 | 11.2 | 10.9 | 8.6 | 7.1 |
| | | | | 100 | 26.19 | 16.3 | 16.3 | 15.4 | 13.8 | 13.4 | 10.6 | 8.8 |
| | | | | 120 | 30.96 | 19.4 | 19.4 | 18.4 | 16.5 | 16.0 | 12.6 | 10.5 |
| | | | | 140 | 36.53 | 23.1 | 23.1 | 21.9 | 19.7 | 19.1 | 15.0 | 12.5 |
| | | | | 160 | 40.49 | 25.8 | 25.8 | 24.4 | 22.0 | 21.3 | 16.8 | 14.0 |
| 450 | 18 | 457 | | 10 | 6.35 | 3.4 | 3.4 | 3.2 | 2.9 | 2.8 | 2.2 | 1.8 |
| | | | | 20 | 7.92 | 4.2 | 4.2 | 4.0 | 3.6 | 3.5 | 2.8 | 2.3 |
| | | | STD | | 9.53 | 5.1 | 5.1 | 4.8 | 4.3 | 4.2 | 3.3 | 2.8 |
| | | | XS | 30 | 11.13 | 6.0 | 6.0 | 5.7 | 5.1 | 4.9 | 3.9 | 3.2 |
| | | | | | 12.7 | 6.8 | 6.8 | 6.5 | 5.8 | 5.6 | 4.4 | 3.7 |
| | | | | 40 | 14.27 | 7.7 | 7.7 | 7.3 | 6.5 | 6.4 | 5.0 | 4.2 |
| | | | | 60 | 19.05 | 10.4 | 10.4 | 9.8 | 8.8 | 8.5 | 6.7 | 5.6 |
| | | | | 80 | 23.83 | 13.1 | 13.1 | 12.3 | 11.1 | 10.8 | 8.5 | 7.1 |
| | | | | 100 | 29.36 | 16.2 | 16.2 | 15.3 | 13.8 | 13.4 | 10.6 | 8.8 |
| | | | | 120 | 34.93 | 19.5 | 19.5 | 18.4 | 16.6 | 16.1 | 12.7 | 10.5 |
| | | | | 140 | 39.67 | 22.3 | 22.3 | 21.1 | 19.0 | 18.4 | 14.5 | 12.0 |
| | | | | 160 | 45.24 | 25.7 | 25.7 | 24.3 | 21.8 | 21.2 | 16.7 | 13.9 |

**Table 20-4** Carbon Steel Grade B Pipes (*continued*)

| Nominal Size | | Outside Diameter (mm) | Schedule | | Wall Thickness (mm) | Temperature (°C) | | | | | | |
|---|---|---|---|---|---|---|---|---|---|---|---|---|
| | | | | | | −29 to +38 | 204 | 260 | 343 | 371 | 399 | 427 |
| | | | | | | Maximum Allowable Stress (MPa) | | | | | | |
| (DN) | (NPS) | | | | | 137.9 | 137.9 | 130.3 | 117.2 | 113.8 | 89.6 | 74.5 |
| 500 | 20 | 508 | | 10 | 6.35 | 3.0 | 3.0 | 2.9 | 2.6 | 2.5 | 2.0 | 1.6 |
| | | | STD | 20 | 9.53 | 4.6 | 4.6 | 4.3 | 3.9 | 3.8 | 3.0 | 2.5 |
| | | | XS | 30 | 12.7 | 6.1 | 6.1 | 5.8 | 5.2 | 5.1 | 4.0 | 3.3 |
| | | | | 40 | 15.09 | 7.3 | 7.3 | 6.9 | 6.2 | 6.0 | 4.8 | 4.0 |
| | | | | 60 | 20.62 | 10.1 | 10.1 | 9.5 | 8.6 | 8.3 | 6.6 | 5.4 |
| | | | | 80 | 26.19 | 12.9 | 12.9 | 12.2 | 11.0 | 10.6 | 8.4 | 7.0 |
| | | | | 100 | 32.54 | 16.2 | 16.2 | 15.3 | 13.8 | 13.4 | 10.5 | 8.7 |
| | | | | 120 | 38.1 | 19.1 | 19.1 | 18.1 | 16.2 | 15.8 | 12.4 | 10.3 |
| | | | | 140 | 44.45 | 22.5 | 22.5 | 21.3 | 19.1 | 18.6 | 14.6 | 12.1 |
| | | | | 160 | 50.01 | 25.5 | 25.5 | 24.1 | 21.7 | 21.0 | 16.6 | 13.8 |
| 600 | 24 | 610 | | 10 | 6.35 | 2.5 | 2.5 | 2.4 | 2.2 | 2.1 | 1.6 | 1.4 |
| | | | STD | 20 | 9.53 | 3.8 | 3.8 | 3.6 | 3.2 | 3.1 | 2.5 | 2.1 |
| | | | XS | | 12.7 | 5.1 | 5.1 | 4.8 | 4.3 | 4.2 | 3.3 | 2.8 |
| | | | | 30 | 14.27 | 5.7 | 5.7 | 5.4 | 4.9 | 4.7 | 3.7 | 3.1 |
| | | | | 40 | 17.48 | 7.1 | 7.1 | 6.7 | 6.0 | 5.8 | 4.6 | 3.8 |
| | | | | 60 | 24.61 | 10.0 | 10.0 | 9.5 | 8.5 | 8.3 | 6.5 | 5.4 |
| | | | | 80 | 30.96 | 12.7 | 12.7 | 12.0 | 10.8 | 10.5 | 8.3 | 6.9 |
| | | | | 100 | 38.89 | 16.1 | 16.1 | 15.2 | 13.7 | 13.3 | 10.5 | 8.7 |
| | | | | 120 | 46.02 | 19.2 | 19.2 | 18.2 | 16.3 | 15.9 | 12.5 | 10.4 |
| | | | | 140 | 52.37 | 22.0 | 22.0 | 20.8 | 18.7 | 18.2 | 14.3 | 11.9 |
| | | | | 160 | 59.54 | 25.3 | 25.3 | 23.9 | 21.5 | 20.9 | 16.4 | 13.7 |

**Table 20-5**　Supply Pipe Dimensions

| Pipe Type | Rated Size, Inches | Actual ID*, Inches | OD†, Inches | Fitting Depth, Inches |
|---|---|---|---|---|
| Galvanized steel | $\frac{1}{8}$ | $\frac{5}{16}$ | $\frac{3}{8}$ | $\frac{1}{4}$ |
| | $\frac{1}{4}$ | $\frac{3}{8}$ | $\frac{1}{2}$ | $\frac{3}{8}$ |
| | $\frac{3}{8}$ | $\frac{1}{2}$ | $\frac{5}{8}$ | $\frac{3}{8}$ |
| | $\frac{1}{2}$ | $\frac{9}{16}$ | $\frac{3}{4}$ | $\frac{1}{2}$ |
| | $\frac{3}{4}$ | $\frac{13}{16}$ | 1 | $\frac{9}{16}$ |
| | 1 | $1\frac{1}{16}$ | $1\frac{1}{4}$ | $\frac{11}{16}$ |
| Copper pipe (type M) | $\frac{1}{2}$ | $\frac{9}{16}$ | $\frac{5}{8}$ | $\frac{1}{2}$ |
| | $\frac{3}{4}$ | $\frac{13}{16}$ | $\frac{7}{8}$ | $\frac{3}{4}$ |
| | 1 | $1\frac{1}{16}$ | $1\frac{3}{16}$ | $\frac{15}{16}$ |
| PVC‡ (STR) | $\frac{1}{2}$ | $\frac{5}{8}$ | $\frac{7}{8}$ | $\frac{1}{2}$ |
| | $\frac{3}{4}$ | $\frac{13}{16}$ | $1\frac{1}{8}$ | $\frac{5}{8}$ |
| | 1 | $1\frac{1}{16}$ | $1\frac{3}{8}$ | $\frac{3}{4}$ |
| CPVC (STR) | $\frac{1}{2}$ | $\frac{1}{2}$ | $\frac{5}{8}$ | $\frac{1}{2}$ |
| | $\frac{3}{4}$ | $\frac{3}{4}$ | 1 | $\frac{5}{8}$ |
| | 1 | 1 | $1\frac{3}{8}$ | $\frac{3}{4}$ |

*ID = inside diameter.
†OD = outside diameter.
‡Dimensions of PVC are different from CPVC dimensions.

**Table 20-6** Various Units, US to UK

| Length | |
|---|---|
| 1 centimeter (cm) | = 10 millimeters (mm) |
| 1 inch | = 2.54 centimeters (cm) |
| 1 foot | = 0.3048 meters (m) |
| 1 foot | = 12 inches |
| 1 yard | = 3 feet |
| 1 meter (m) | = 100 centimeters (cm) |
| 1 meter (m) | = 3.280839895 feet |
| 1 furlong | = 660 feet |
| 1 kilometer (km) | = 1000 meters (m) |
| 1 kilometer (km) | = 0.62137119 miles |
| 1 mile | = 5280 ft |
| 1 mile | = 1.609344 kilometers (km) |
| 1 nautical mile | = 1.852 kilometers (km) |

| Area | |
|---|---|
| 1 square foot | = 144 square inches |
| 1 square foot | = 929.0304 square centimeters |
| 1 square yard | = 9 square feet |
| 1 square meter | @ 10.7639104 square feet |
| 1 acre | = 43,560 square feet |
| 1 hectare | = 10,000 square meters |
| 1 hectare | @ 2.4710538 acres |
| 1 square kilometer | = 100 hectares |
| 1 square mile | @ 2.58998811 square kilometers |
| 1 square mile | = 640 acres |

| Speed | |
|---|---|
| 1 mile per hour (mph) | @ 1.46666667 feet per second (fps) |
| 1 mile per hour (mph) | = 1.609344 kilometers per hour |
| 1 knot | @ 1.150779448 miles per hour |
| 1 foot per second | @ 0.68181818 miles per hour (mph) |
| 1 kilometer per hour | @ 0.62137119 miles per hour (mph) |

| Volume | |
|---|---|
| 1 US tablespoon | = 3 US teaspoons |
| 1 US fluid ounce | @ 29.57353 milliliters (ml} |
| 1 US cup | = 16 US tablespoons |
| 1 US cup | = 8 US fluid ounces |
| 1 US pint | = 2 US cups |
| 1 US pint | = 16 US fluid ounces |
| 1 liter (l) | @ 33.8140227 US fluid ounces |
| 1 liter (l) | = 1000 milliliters (ml) |
| 1 US quart | = 2 US pints |
| 1 US gallon | = 4 US quarts |
| 1 US gallon | = 3.78541178 liters |

| Weight | |
|---|---|
| 1 milligram (mg) | = 0.001 grams (g) |
| 1 gram (g) | = 0.001 kilograms (kg) |
| 1 gram (g) | = 0.035273962 ounces |
| 1 ounce | = 28.34952312 grams (g) |
| 1 ounce | = 0.0625 pounds |
| 1 pound (lb) | = 16 ounces |
| 1 pound {lb) | = 0.45359237 kilograms (kg) |
| 1 kilogram (kg) | = 1000 grams |
| 1 kilogram (kg) | = 35.273962 ounces |
| 1 kilogram (kg) | = 2.20462262 pounds (lb) |
| 1 stone | = 14 pounds |
| 1 short ton | = 2000 pounds |
| 1 metric ton | = 1000 kilograms (kg) |

**Temperature**

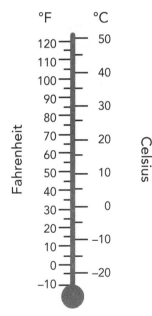

## Review Questions

1. What is the system of measures used in Europe called?

   _____

2. What is the system of measures used in the U.S. called?

   _____

3. What is meant by conversion?

   _____

4. What countries are included in the term UK (United Kingdom)?

   _____

5. When did the push for converting the world to using the metric system occur?

   _____

# 21 | Conversions and Equivalents

## Performance Objectives

After studying this chapter, you will:

- Know more about conversion factors.

- Know more about unit prefixes.

- Be able to convert the inch to metric equivalents (two place decimals).

- Be able to convert U.S. gallons to liters.

- Be able to convert liters to U.S. gallons.

- Be able to convert fractional inches to millimeters.

- Be able to convert millimeters to decimal inches.

- Know more about decimal equivalents.

- Be able to convert English units to metric units.

- Be able to answer the review questions at the end of this chapter.

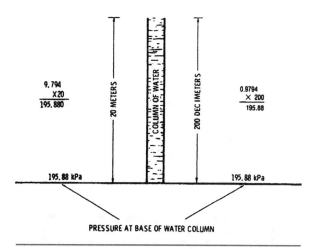

**Figure 21-1**  Converting temperatures.

Formulas can be used for converting temperature given in one scale to that of the other. For instance, when using a calculator it is easier to compute. See Figure 21-1.

## Temperature

$$F = 1.8C + 32 \text{ or } 1.8 \times C + 32$$

$$C = F - 32 \div 1.8$$

## Pressure

Kilopascal (kPa) is the unit recommended for fluid pressure for almost all fields of use, such as barometric pressure, gas pressure, tire pressure, and water pressure.

Atmospheric pressure is: 101 kPa metric, and 14.7 psi English; 6.894757 kPa = 1psi.

To find head pressure in decimeters when pressure is given in kilopascal (kPa), divide the pressure by 0.9794.

To find pressure in kPa of a column of water given in decimeters, multiply decimeters by 0.9794.

To find head pressure in meters when pressure is given in kilopascals, divide pressure by 9.794. To find pressure in kPa of a column water given in meters, multiply meters by 9.794.

## Additional Valuable Metric Information

The amount of heat required to change ice to liquid water is 144 Stu's per lb. (335 joules per kilogram). Then, generally speaking, 6 cm of mercury = 8 kPa pressure and one cubic meter of air weighs1.214 kilograms (kg).

Atmospheric pressure of 101.3 kPa will balance or support a column of mercury 76 cm high. When you know inches of mercury (in Hg), you can multiply this quantity by 3.386389 to find the number of kilopascals (kPa). A column of water is 9.794 kPa per meter or 0.2476985 kPa per inch.

If you take one milliliter (ml) of water, it has a mass of 1 gram, and 1 foot-lb = 1.3558 newton-meter (Nm) (bending moment of torque).

By studying the following information about metrics and the English systems of measurement, you will be able to work in the plumbing trade more efficiently or in any position requiring a knowledge of and an ability to convert various units of measurement. Examine the tables on the next few pages to help you learn how to quickly quote measurements in either system of measurement. Many manufacturers of pipes have the technical information on the pipes in catalogs and serve a wider group of nations if both measurements are given. Table 21-1 assists you in converting English measuremetns to metric measurements. Table 21-2 provides you with metric abbreviations. Table 21-3 shows length and liquid volume values, and quick reference to the chart provides both metric and English equivalents. Table 21-4 provides information on common temperatures as well as inches and feet to meters. Table 21-5 provides parts of an inch converted to millimeters and inches and parts of an inch to centimeters. Other tables are provided to cover all aspects of the metric measurement system.

**Table 21-1**   English to Metric Conversion

| Symbol | To convert from | Multiply by | To determine | Symbol |
|---|---|---|---|---|
| **LENGTH** | | | | |
| IN | inch | 25.4 | millimeters | mm |
| FT | feet | 0.3048 | meters | m |
| YD | yards | 0.9144 | meters | m |
| MI | miles | 1.609344 | kilometers | km |
| **AREA** | | | | |
| SI | square inches | 645.16 | square millimeters | mm² |
| SF | square feet | 0.09290304 | square meters | m² |
| SY | square yards | 0.83612736 | square meters | mi |
| A | acres | 0.4046856 | hectares | ha |
| MI² | square miles | 2.59 | square kilometers | km² |
| **VOLUME** | | | | |
| CI | cubic inches | 16.387064 | cubic centimeters | cm³ |
| CF | cubic feet | 0.0283168 | cubic meters | m³ |
| CY | cubic yards | 0.764555 | cubic meters | m³ |
| GAL | gallons | 3.78541 | liters | L |
| OZ | fluid ounces | 0.0295735 | liters | L |
| MBM | thousand feet board | 2.35974 | cubic meters | m³ |
| **MASS** | | | | |
| LB | pounds | 0.4535924 | kilograms | kg |
| TON | short tons (2000 lbs) | 0.9071848 | metric tons | t |
| **PRESSURE AND STRESS** | | | | |
| PSF | pounds per square foot | 47.8803 | pascals | Pa |
| PSI | pounds per square inch | 6.89476 | kilopascals | kPa |
| PSI | pounds per square inch | 0.00689476 | megapascals | Mpa |
| **DISCHARGE** | | | | |
| CFS | cubic feet per second | 0.02831 | cubic meters per second | m³/s |
| **VELOCITY** | | | | |
| FT/SEC | feet per second | 0.3048 | meters per second | m/s |
| **INTENSITY** | | | | |
| IN/HR | inch per hour | 25.4 | millimeters per hour | mm/hr |
| **FORCE** | | | | |
| LB | pound (force) | 4.448222 | newtons | N |
| **POWER** | | | | |
| HP | horsepower | 746.0 | watts | w |
| **TEMPERATURE** | | | | |
| °F | degrees Fahrenheit | 5 x (°F – 32)/9 | degrees Celsius | °C |
| **DENSITY** | | | | |
| lb/ft³ | pounds per cubic foot | 16.01846 | kilograms per cubic meter | kg/m³ |
| **ACCELERATION** | | | | |
| G | freefall, standard | 9.807 | meters per second squared | m/s² |

To convert from Metric to English, divide by the above conversion factors.

**Table 21-2**   Metric Abbreviations

| | |
|---|---|
| ANMC | American National Metric Council |
| ANSI | American National Standards Institute |
| NBS | National Bureau of Standards |
| SI | International System of Units |
| m | meter |
| km | kilometer |
| cm | centimeter |
| mm | millimeter |
| kg | kilogram |
| g | gram |
| mg | milligram |
| l | liter |
| ml | milliliter |
| Nm | newton-meter |
| °C | degrees celsius |
| kPa | kilopascal |
| J | joule |
| inHg | inches of mercury |
| Pa | pascal |

**Table 21-3**   Length and Volume Conversions

| Length | | |
|---|---|---|
| **Metric** | | **English** |
| 1 meter | = | 39.37 inches |
| 1 meter | = | 1000 millimeters |
| 1 meter | = | 100 centimeters |
| 1 meter | = | 10 decimeters |
| 1 kilometer | = | 0.625 miles |
| 1.609 kilometers | = | 1 mile |
| 25.4 millimeters | = | 1 inch |
| 2.54 centimeters | = | 1 inch |
| 304.8 millimeters | = | 1 foot |
| 1 millimeter | = | 0.03937 inches |
| 1 centimeter | = | 0.3937 inches |
| 1 decimeter | = | 3.937 inches |
| **Volume—Liquid** | | |
| **Metric** | | **English** |
| 3.7854 liters | = | 1 gallon |
| 0.946 liters | = | 1 quart |
| 0.473 liters | = | 1 pint |
| 1 liter | = | .264 gals. or 1.05668 qts. |
| 1 liter | = | 33.814 ounces |
| 29.576 milliliters | = | 1 fluid ounce |
| 236.584 milliliters | = | 1 cup |

Note: 1 liter contains 1000 milliliters

**Table 21-4**   Inches and Feet to Millimeters

| Inches to Millimeters | | Feet to Millimeters | |
|---|---|---|---|
| English (inches) | Metric (mm) | English (feet) | Metric (mm) |
| 1 | 25.4 | 2 | 609.6 |
| 2 | 50.8 | 3 | 914.4 |
| 3 | 76.2 | 4 | 1219.2 |
| 4 | 101.6 | 5 | 1524.0 |
| 5 | 127 | 6 | 1828.8 |
| 6 | 142.5 | 7 | 2133.6 |
| 7 | 177.8 | 8 | 2438.4 |
| 8 | 203.2 | 9 | 2743.2 |
| 9 | 228.6 | 10 | 3048.0 |
| 10 | 254 | 20 | 6096.0 |
| 11 | 279.4 | | |
| 12 | 304.8 | | |

Note: Round off to nearest millimeter. Thus,
4" = 102 mm
3" = 76 mm
2" = 51 mm, and so on.

**Table 21-5**  Parts of an Inch to Millimeters and Inches and Parts of an Inch to Centimeters

Parts of an Inch to Millimeters

| English (parts of an inch) | Metric (mm) | English (parts of an Inch) | Metric (mm) |
|---|---|---|---|
| 1/32 | 0.79375 (0.80) | 9/16 | 14.2875 (14.3) |
| 1/16 | 1.5875 (1.6) | 5/8 | 15.8750 (15.9) |
| 1/8 | 3.175 (3.2) | 11/16 | 17.4625 (17.5) |
| 3/16 | 4.7625 (4.8) | 3/4 | 19.0500 (19.1) |
| 1/4 | 6.35 (6.4) | 13/16 | 20.6375 (20.6) |
| 5/16 | 7.9375 (7.9) | 7/8 | 22.2250 (22.2) |
| 3/8 | 0.5250 (9.5) | 15/16 | 23.8175 (23.8) |
| 7/16 | 11.1125 (11.1) | 1 | 25.4000 (25.4) |
| 1/2 | 12.7000 (12.7) | | |

Note: In most cases it is best to round off to the nearest millimeter. Thus:

17.4625 would be: 17 mm

20.6375 would be: 21 mm

Inches and Parts of an Inch to Centimeters

| Inches (parts of an inch) | Metric (cm) | Inches (parts of an Inch) | Metric (cm) |
|---|---|---|---|
| 1/16 | 0.15875 | 2½ | 6.35 |
| 1/8 | 0.3175 | 3 | 7.62 |
| 1/4 | 0.635 | 4 | 10.16 |
| 3/8 | 0.9525 | 5 | 12.70 |
| 1/2 | 1.27 | 6 | 15.24 |
| 5/8 | 1.5875 | 7 | 17.78 |
| 3/4 | 1.905 | 8 | 20.32 |
| 7/8 | 2.2225 | 9 | 22.86 |
| 1 | 2.54 | 10 | 25.40 |
| 1¼ | 3.175 | 11 | 27.94 |
| 1½ | 3.81 | 12 | 30.48 |
| 2 | 5.08 | | |

**Table 21-6**   Metric Conversion Chart: English to Metric

| English | | Metric |
|---|---|---|
| Inches (in) | x | 2.54 = centimeters |
| Feet (ft) | x | .3 =meters |
| Yards (yd) | x | .9 =meters |
| Miles (mi) | x | 1.6 = kilometers |
| Square inches (in²) | x | 6~5 = square centimeters |
| Square feet (ft²) | x | .1 = square meters |
| Square yards (yd²) | x | .8 = square meters |
| Acres | x | .4 = hectares |
| Cubic feet (ft³) | x | .03 = cubic meters |
| Cords (cd) | x | 3.6 = cubic meters |
| Quarts (lq) (qt) | x | .9 = liters |
| Gallons (gal) | x | .004 = liters |
| Ounces (avdp) (oz) | x | 28. 4 = grams |
| Pounds (avdp) (lb) | x | .5 = kilograms |
| Horsepower (hp) | x | .7 =kilowatts |

**Table 21-7**   Metric Conversion Chart: Metric to English

| Metric | | English |
|---|---|---|
| Centimeters (cm) | x | .39 = inches |
| Meters (m) | x | 3.3 = feet |
| Meters (m) | x | 1.1 = yards |
| Kilometers (km) | x | .6 = miles |
| Sq. centimeters (cm²) | x | .2 = square inches |
| Square meters (m²) | x | 10.8 = square feet |
| Square meters (m²) | x | 1.2 = square yards |
| Hectares (ha) | x | 2.5 = acres |
| Cubic meters (m³) | x | 35.3 = cubic feet |
| Liters (1) | x | 1.1 = quarts (lq) |
| Cubic meters (m³) | x | 284.2 = gallons |
| Grams (g) | x | .04 = ounces (avdp) |
| Kilograms (kg) | x | 2.2 = pounds (avdp) |
| Kilowatts (kW) | x | 1.3 = horsepower |

**Table 21-8**   Units of Length and Measure and Temperature Conversions

Length
```
12 inches . . . . . . . . . . . . . . . . . . . . . . . . . . . . . . . . . . . .1 foot
36 inches or 3 feet . . . . . . . . . . . . . . . . . . . . . . . . . . . 1 yard
1760 yards or 5280 feet. . . . . . . . . . . . . . . . . . . . . . . .1 mile
```

Liquid Measure
```
8 ounces. . . . . . . . . . . . . . . . . . . . . . . . . . . . . . . . . . . . . 1 cup
16 ounces or 2 cups. . . . . . . . . . . . . . . . . . . . . . . . . . . .1 pint
32 ounces or 4 cups or 2 pints. . . . . . . . . . . . . . . . . . .1 quart
64 ounces or 4 pints or 2 quarts . . . . . . . . . . . . . . . 1/2 gallon
128 ounces or 16 cups or 8 pints or 4 quarts . . . . . . . 1 gallon
```

Temperature Conversions

From Fahrenheit to Centigrade
To convert from degrees Fahrenheit to degrees Centigrade, subtract 32 degrees from the temperature and multiply by 5/9:

| Fahrenheit | 0 | 10 | 20 | 30 | 40 | 50 | 60 | 70 | 80 | 90 | 100 |
|---|---|---|---|---|---|---|---|---|---|---|---|
| Centigrade | −18 | −12 | −7 | −1 | 4 | 10 | 16 | 21 | 27 | 32 | 38 |

From Centigrade to Fahrenheit
To convert from degrees Centigrade to degrees Fahrenheit, multiply the temperature by 1.8 and add 32 degrees:

| Centigrade | −10 | −5 | 0 | 5 | 10 | 15 | 20 | 25 | 30 | 35 | 40 |
|---|---|---|---|---|---|---|---|---|---|---|---|
| Fahrenheit | 14 | 23 | 32 | 41 | 50 | 59 | 68 | 77 | 86 | 95 | 104 |

**Table 21-9**   Fractions and Decimals

| | |
|---|---|
| $\frac{1}{64}$ = .015625 | $\frac{33}{64}$ = .515625 |
| $\frac{1}{32}$ = .03125 | $\frac{17}{32}$ = .53125 |
| $\frac{3}{64}$ = .046875 | $\frac{35}{64}$ = .546875 |
| $\frac{1}{16}$ = .0626 | $\frac{9}{16}$ = .5625 |
| $\frac{5}{64}$ = .078125 | $\frac{37}{64}$ = .578125 |
| $\frac{3}{32}$ = .09375 | $\frac{19}{32}$ = .59375 |
| $\frac{7}{64}$ = .109375 | $\frac{39}{64}$ = .609375 |
| $\frac{1}{8}$ = .125 | $\frac{5}{8}$ = .625 |
| $\frac{8}{64}$ = .140625 | $\frac{41}{64}$ = .640625 |
| $\frac{5}{32}$ = .15625 | $\frac{21}{32}$ = .65625 |
| $\frac{11}{64}$ = .171875 | $\frac{43}{64}$ = .671875 |
| $\frac{3}{16}$ = .1875 | $\frac{11}{16}$ = .6875 |
| $\frac{12}{64}$ = .203125 | $\frac{45}{64}$ = .703125 |
| $\frac{7}{32}$ = .21875 | $\frac{23}{32}$ = .71875 |
| $\frac{15}{64}$ = .234375 | $\frac{47}{64}$ = .734375 |
| $\frac{1}{4}$ = .25 | $\frac{3}{4}$ = .75 |
| $\frac{17}{64}$ = .265625 | $\frac{49}{64}$ = .765625 |
| $\frac{9}{32}$ = .28125 | $\frac{25}{32}$ = .78125 |
| $\frac{19}{64}$ = .296875 | $\frac{51}{64}$ = .796875 |
| $\frac{5}{16}$ = .3125 | $\frac{13}{16}$ = .8125 |
| $\frac{21}{64}$ = .328125 | $\frac{53}{64}$ = .828125 |
| $\frac{11}{32}$ = .34375 | $\frac{27}{32}$ = .84375 |
| $\frac{23}{64}$ = .359375 | $\frac{55}{64}$ = .859375 |
| $\frac{3}{8}$ = .375 | $\frac{7}{8}$ = .875 |
| $\frac{25}{64}$ = .390625 | $\frac{57}{64}$ = .890625 |
| $\frac{13}{32}$ = .40625 | $\frac{29}{32}$ = .90625 |
| $\frac{27}{64}$ = .421875 | $\frac{59}{64}$ = .921875 |
| $\frac{7}{16}$ = .4375 | $\frac{15}{16}$ = .9375 |
| $\frac{29}{64}$ = .453125 | $\frac{61}{64}$ = .953125 |
| $\frac{15}{32}$ = .46875 | $\frac{31}{32}$ = .96875 |
| $\frac{31}{64}$ = .484375 | $\frac{63}{64}$ = .984375 |
| $\frac{1}{2}$ = .5 | |

## Review Questions

1. What does the metric unit *millimeter* mean?

   _____

2. What does the metric system use for measuring *pressure?*

   _____

3. How do you convert Centigrade to Fahrenheit?

   _____

4. Where do you find the metric system of measurement used as standard for all measurements?

   _____

5. Convert 110°F to Celsius.

   _____

6. Whose name is used for the SI system most often?

   _____

7. Convert 72°F to Celsius.

   _____

8. Convert –10°F to Celsius.

   _____

9. Convert 100°C to Fahrenheit.

   _____

10. Convert 304.8 mm to U.S. or UK measurement units.

    _____

# 22 Fractions, Angles, Shapes and Forms

## Performance Objectives

After studying this chapter, you will:

- Understand where the basic units in math originated and how.

- Know more about metric values and how to convert them.

- Know what a mole is and what it measures.

- Know how time is measured.

- Understand the importance of knowing about triangles and circles.

- Be able to convert fractions to metric and metric fractions to U.S. units.

- Be able to answer the review questions at the end of this chapter.

# Math

Fractions, angles and other shapes and forms are included in the study of plumbing. As you know, plumbing encompasses a wide variety of pipes, drains and vents as well as other devices that are made or installed by the magic of measurement. Plumbers must be familiar with various units of measurement.

When working with the large number of installations required by today's plumbing, mathematics, which governs their construction and placement, must be used. One illustration of math being used to solve a plumbing problem is shown in Figures 22-1 and 22-2, in which a pipe connection of 45 degrees is made with elbows. Of course, charts and tables have been provided for most of the problems, making it more efficient for the plumber so that he can concentrate on the demands of the plumbing trade rather than spending time figuring out various angles, volumes and linear measures.

The international standard of measurement is a coherent system of measurement based on seven basic units. They are all based on natural units of measurement—units which are constant, such as the resonance of atoms, the wavelength of light, the speed of light, the mass of objects, the freezing and boiling points of water, and so forth.

# The Meter

The basic unit of length is the *meter* in the metric system and the *foot* in the U.S. or English system. The meter is technically defined as 1,650,763.73 wavelengths in a vacuum of orange-red line of the spectrum of Krypton-86. All other division of length, such as millimeters, centimeters, and kilometers are obtained by multiplying or dividing this unit by 10s.

Mass is the resistance to change in motion. Under the earth's gravitational pull, mass is measured as weight. The basic unit of mass is the kilogram (in the U.S. system it is the pound). Common units used are the gram, kilogram, and tonnes. A newton is the force applied to a 1 kilogram mass to accelerate the mass at 1 meter per second per second.

The basic unit of time is the second (in both the metric system and the U.S. system). This unit was originally set by that period of time involved in the rotation of the earth using the old English unit of two twelve-hour periods. One second of time is derived by calculating $\frac{1}{60} \times \frac{1}{60} \times \frac{1}{24} = \frac{1}{86,4000}$ of a rotation (the mean solar day) of the earth. We now have atomic clocks that are very accurate. Today, time is based on the half-life of uranium to lead.

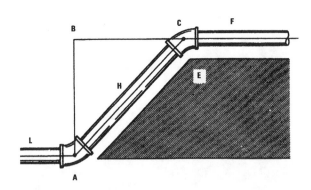

**Figure 22-1** Pipe connected by two 45-degree elbows.

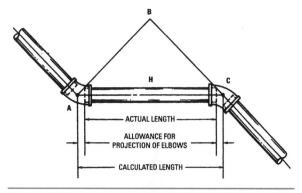

**Figure 22-2** Another method for finding length of H.

## Other Units of Measurement

Other writs of measure include the mole, a basic unit of the amount of a substance.

The candela is the unit of luminous intensity, or 1 candela radiates a light of 4 lumens.

Table 22-1 shows the derived units of the SI (Systeme International d'Unites) or metric system. Various combinations of angles are used to design manufactured products.

Angular dimensions are measured from a reference. The reference may be one leg of the angle being measured. A line that is being rotated about one end will generate angles when referred to the starting position. If the line moves $\frac{1}{8}$ of a rotation, the angle generated will be 45°. If rotated $\frac{1}{4}$ of a rotation the angle will be 90°. When a complete revolution is made, the angle is 360° and a circle will have been generated. The 90° angle (square, or right angle) is the most common angle used and serves as a reference angle. Figure 22-1 shows how the triangle is formed by the offsetting of a pipe.

## Circle

The Greek letter $\pi$ (called pi) is used to represent 3.14159, the circumference of a circle whose diameter is 1. The circumference of a circle equals the diameter multiplied by 3.14159. The reason for using the decimal 0.7854 to calculate the area of a circle is shown in Figure 22-2. For example: if the distance between pipelines L and F in Figure 22-2 is 20 inches (offset AB), what length of pipe (H) is required to connect elbows A and C? When 45-degree elbows are used, both offsets are equal. Thus, substituting in the equation: make sure to put in square root radical over the 20 + 20.

$$AC = \sqrt{20^2 + 20^2} = \sqrt{800} = 28.28 \text{ inches}$$

The length of pipe just calculated does not allow for the projections of the elbows. This must be taken into account, as can be seen in Figure 22-2.

## Triangle

Most of the methods used for calculating pipe offset changes use the triangle. Figure 22-2 shows a pipe change of direction where it is necessary to change the position of the pipeline L to a parallel position F to avoid some obstruction, such as a wall E. When the two lines L and F are to be fitted with elbows having an angle of 90°, the plumber must find the distance of the pipe H connecting the two elbows A and C. The distance BC must also be determined in order to fix the point A so that elbows A and C will be in alignment. There are several methods of solving this problem, of which two follow:

$$C^2 = (AB)^2 + (BC)^2 \text{ from which}$$
$$AC = \sqrt{AB^2 + BC^1}$$

Example: If the distance between pipelines L and F in Figure 22-1 is 20 inches (offset AB), what length of pipe (H) is required to connect elbows A and C? When 45° elbows are used, both offsets are equal. Thus, substituting in the equation:

$$AC = \sqrt{20^2 + 20^2} = \sqrt{800} = 28.28 \text{ inches}$$

The length of the pipe just calculated does not allow for the projections of the elbows. This must be taken into account, as shown in Figure 22-2.

For method 2 the following rule will be found convenient in determining the length of pipe between 45° elbows.

Rule: For each inch of offset, add $\frac{53}{128}$ of an inch, and the result will be the length between centers of the elbows.

Example: Calculate the length AC (Fig. 22-1) by the preceding rule:

$$20 \times {}^{53}\!/_{128} = {}^{1060}\!/_{128} = 8.28125$$

Now add this to the offset: $20 + 8.28125 = 28.28125$.

# NOTE

Take 9 and divide it by 32 and your calculator will produce 0.28125, which means the answer is really **28.28**, just like the first example.

This is the calculated length; to obtain the actual length, deduct the allowance for the projection of the elbows, as in Figure 22-2.

# Quickly Accessible Information

Derived units from the SI system of measurement (Metric), some common symbols, common abbreviations and mathematical definitions are included here for quick reference. A table for quickly arriving at square roots, cube roots and reciprocals is included for your daily encounters on the job.

Other tables cover such things as boiling points of water at various pressures above atmospheric, and decimal equivalents so you can use the fractional information in a form that the computer or calculator will accept.

Some more useful information is also included for your updating and reference in case you have forgotten.

**Table 22-1A**   Metric to English for Liquids and Length

| Liquid Volume | |
|---|---|
| **Metric** | **English** |
| 3.7854 L | 1 gallon |
| 0.946 L | 1 quart |
| 0.743 L | 1 pint |
| 1 L | 0.264 gallons or 1.05668 quarts |
| 1 L | 33.814 ounces |
| 29.576 mm | 1 fluid ounce |
| 236.584 mm | 1 cup |

Note: 1 liter contains 1,000 milliliters.

| Length | |
|---|---|
| **Metric** | **English** |
| 1 m | 39.37 inches |
| 1 m | 1000 millimeters |
| 1 m | 100 centimeters |
| 1 m | 10 decimeters |
| 1 km | 0.625 miles |
| 1.609 km | 1 mile |
| 25.4 mm | 1 inch |
| 2.54 cm | 1 inch |
| 304.8 mm | 1 foot |
| 1 mm | 0.03937 inch |
| 1 cm | 0.3937 inch |
| 1 dm | 3.937 inches |

**Table 22-1B**  U.S. Weights and Measures

| Mile Measurements | | | | | |
|---|---|---|---|---|---|
| 1 Statute Mile | = | 5,280 feet | | | |
| 1 Scots Mile | = | 5,952 feet | | | |
| 1 Irish Mile | = | 6,720 feet | | | |
| 1 Russian Verst | = | 3,504 feet | | | |
| 1 Italian Mile | = | 4,401 feet | | | |
| 1 Spanish Mile | = | 15,084 feet | | | |
| **Other Linear Measurements** | | | | | |
| 1 Hand | = | 4 inches | 1 Fathom | = | 6 feet |
| 1 Span | = | 9 inches | 1 Furlong | = | 10 Chains |
| 1 Chain | = | 22 yards | 1 Cable | = | 608 feet |
| 1 Link | = | 7.92 inches | | | |
| **Square Measure** | | | | | |
| 144 Square Inches | = | 1 Square Foot | | | |
| 9 Square Feet | = | 1 Square Yard | | | |
| 301/4 Square Yards | = | 1 Square Rod | | | |
| 40 Rods | = | 1 Acre | | | |
| 640 Acres | = | 1 Square Mile | | | |
| 1 Square Mile | = | 1 Section | | | |
| 36 Sections | = | 1 Township | | | |
| **Cubic of Solid Measure** | | | | | |
| 1 Cubic Foot | = | 1,728 Cubic Inches | | | |
| 1 Cubic Yard | = | 27 Cubic Feet | | | |
| 1 Cubic Foot | = | 7.48 Gallons | | | |
| 1 Gallon (Water) | = | 8.34 Pounds | | | |
| 1 Gallon (U.S.) | = | 231 Cubic Inches of Water | | | |
| 1 Gallon (Imperial) | = | $277\frac{1}{4}$ Cubic Inches of Water | | | |
| **Linear Measure** | | | | | |
| | | 1 Inch | = | 2.540 Centimeters | |
| 12 Inches | = | 1 Foot | = | 3.048 Decimeters | |
| 3 Feet | = | 1 Yard | = | 9.144 Decimeters | |
| 5.5 Yards | = | 1 Rod | = | 5.029 Meters | |
| 40 Rods | = | 1 Furlong | = | 2.018 Hectometers | |
| 8 Furlongs | = | 1 Mile | = | 1.609 Kilometers | |

**Table 22-1C**   Working with the Metric System

| (The unit is the meter = 39.37 inches) | | | | |
|---|---|---|---|---|
| 1 Cubic Centimeter | = | 1,000 Cubic Millimeters | = | 0.3102 Cubic Inches |
| 1 Cubic Decimeter | = | 1,000 Cubic Centimeters | = | 61.02374 Cubic Inches |
| 1 Cubic Meter | = | 1,000 Cubic Decimeters | = | 35.31467 Cubic Feet |
| | = | 1 Stere | = | 1.30795 Cubic Yards |
| 1 Cubic Cenimeter (Water) | | | = | 1 Gram |
| 1,000 Cubic Centimeters (Water) | = | 1 Liter | = | 1 Kilogram |
| 1 Cubic Meter (1,000 Liters) | | | = | 1 Metric Ton |
| **Measure of Weight** | | | | |
| (The unit is the gram = 0.035274 ounces) | | | | |
| 1 Milligram | | | = | 0.015432 Grains |
| 1 Centigram | = | 10 Milligrams | = | 0.15432 Grains |
| 1 Decigram | = | 10 Centigrams | = | 1.5432 Grains |
| 1 Gram | = | 10 Decigrams | = | 15.432 Grains |
| 1 Dekagram | = | 10 Grams | = | 3.5274 Ounces |
| 1 Kilogram | = | 10 Hectograms | = | 2.2046223 Pounds |
| 1 Myriagram | = | 10 Kilograms | = | 22.046223 Pounds |
| 1 Quintal | = | 10 Myriagrams | = | 1.986412 CWT |
| 1 Metric Ton | = | 10 Quintal | = | 22,045.622 Pounds |
| 1 Dram | = | 1.77186 Grams | | |
| | = | 27.3438 Grains | | |
| 1 Metric Ton | = | 2,204.6223 Pounds | | |
| **Measures of Capacity** | | | | |
| (The unit is the liter = 1.0567 liquid quarts) | | | | |
| 1 Centiliter | = | 10 Milliliters | = | 0.338 Fluid Ounces |
| 1 Deciliter | = | 10 Centiliters | = | 3.38 Fluid Ounces |
| 1 Liter | = | 10 Deciliters | = | 33.8 Fluid Ounces |
| 1 Dekaliter | = | 10 Liters | = | 0.284 Bushel |
| 1 Hectoliter | = | 10 Dekaliters | = | 2.84 Bushes |
| 1 Kiloliter | = | 10 Hectoliters | = | 264.2 Gallons |

Note: $\frac{Kilometers}{8} \times 5 = Miles$ or $\frac{Miles}{5} \times 8 = Kilometers$

**Table 22-2** Symbols for Plumbing Fixtures

| Pipe | | Fixtures | |
|---|---|---|---|
| Cold water | — · — · — · — · — | Bathtub | |
| Filtered cold water | —FCW—FCW—FCW— | Bidet | |
| Hot water | — ·· — ·· — ·· — | Double compartment sink | |
| Hot water return | — — — — — | Handicap lavatory | |
| Drain | —D—D—D— | Single bowl sink | |
| Drinking water supply | —DWS—DWS—DWS— | Urinal | |
| Drinking water return | —DWR—DWR—DWR— | Wall-mounted water closet | |
| Natural gas | —NG—NG—NG— | Shower head | SH |
| **Valves** | | Shower stall | |
| Gate valve | | Hot water tank | HWT |
| Globe valve | | Drinking fountain | DF |
| Check valve | | Floor drain, round | FD |
| Angle globe valve | | Floor drain square | FD |
| Butterfly valve | | **Fittings** | |
| Ball valve | | 90° elbow | |
| Modulating control valve | | 45° elbow | |
| Two-position control valve | | Tee | |
| Motorized valve | | Wye | |
| Flanged valve | | Union | |
| Manual air vent | | Cap | |
| Automatic air vent | | Bushing | |

Data from the U.S. Department of Veteran Affairs, TIL—Standard Details (PG-18-4) (http://www.cfm.va.gov/til.sDetail.asp#22) Accessed July 13, 2010; Plumbing pipe and fitting symbols used with permission of Plumbing Help (http://www.plumbinghelp.ca/articles_plumbing_symbols.php).

**Table 22-3**   Commonly Used Abbreviations

| Abbreviation | Description |
|---|---|
| A or a | Area |
| Atm. Pres. | Standard atmospheric pressure |
| av. | Average |
| AWG | American wire gauge |
| bbl | Barrels |
| B or b | Bredth |
| bhp | Brake horsepower |
| BM | Board measure |
| BTU | British thermal units |
| BWG | Birmingham wire gauge |
| C of g | Center of gravity |
| cal. val. | Calorific value |
| cm | Centimeters |
| Cp | Specific heat at constant pressure |
| Cv | Specific heat at constant volume |
| cu. | Cubic |
| cyl. | Cylinder |
| D or d | Depth or diameter |
| deg. | Degrees |
| diam. | Diameter |
| evap. | Evaporation |
| F | Fahrenheit |
| g. | Gravity acceleration |
| gals. | Gallons |
| gpm | Gallons per minute |
| H or h | Height or head of water |
| hp | Horsepower |
| ihp | Indicated horsepower |
| kg | Kilograms |
| lb | Pounds |
| lb-ft | Pound-feet |
| lb per sq. in. | Powers per square inch |
| log | Logarithm to the base 10 |
| loge or ln | Logarithm to the base of e |
| min. | Minute |
| mm Hg | Millimeters of mercury (pressure) |

**Table 22-4**   Metric Abbreviations

| Abbreviation | Meaning |
|---|---|
| cm | Centimeter |
| cm$^2$ | Square centimeter |
| cm$^3$ | Cubic centimeter |
| dm | Decimeter |
| dm$^2$ | Square decimeter |
| dm$^3$ | Cubic decimeter |
| g | Grams |
| inHg | Inches of mercury |
| J | Joule |
| kg | Kilogram |
| km | Kilometer |
| km2 | Square kilometer |
| kPa | Kilopascal |
| L | Liter |
| m | Meter |
| m$^2$ | Square meter |
| mm$^3$ | Cubic millimeter |
| mg | Milligram |
| mL | Milliliter |
| mm | Millimeter |
| NBS | National Bureau of Standards |
| N-m | Newton-meter |
| °C | Degrees Celsius |
| Pa | Pascal |
| SI | International System of Units |

**Table 22-5**   Mathematical Symbols

| Meaning | Description | Symbol |
|---|---|---|
| Equals | Both sides of an equation have the same value | = |
| Minus | One number is subtracted from another number. | − |
| Plus | One number is added to another number. | + |
| Multiplied by | One number is multiplied by another number. | × |
| Divided by | One number is divided by another number. | ÷ |
| Exponential power | x, a number, is multiplied by itself a (also a number) times. | $x^a$ |
| Square root | When a number is multiplied by itself once, the result is the number squared (e.g., 2 × 2 = 4, or 2$^2$). The number being multiplied by itself is the square root of the product (e.g., 2 is the square root of 4). | $\sqrt{\phantom{x}}$ |

**Table 22-6** Square and Cube Measurements

| Measurement | Equivalent |
|---|---|
| 2.59km² | 1 square mile |
| 0.093 m² | 1 square foot |
| 6.451 cm² | 1 square inch |
| 0.765 m³ | 1 cubic yard |
| 0.028316 m³ | 1 cubic foot |
| 16.387 cm³ | 1 cubic inch |
| 1 m³ | 35.3146 cubic feet |
| 929.03 cm² | 1 square foot |
| 10,000 cm² | 1 m² |
| 1 m³ | 1,000,000 cm³ or 1,000 dm³ |
| 10.2 cm of water | 1 kPa of pressure |
| 51 cm of water | 5 kPa |
| 1 m of water | 9.8 kPa |
| 1 m³ of air | 1.214 kg |
| 1 cubic foot | 28,316.846522 cm² |

**Table 22-7** Fractional-Decimal Equations

| Fraction | Decimal |
|---|---|
| $\frac{1}{16}$ inch | 0.06 |
| $\frac{1}{8}$ inch | 0.13 |
| $\frac{3}{16}$ inch | 0.19 |
| $\frac{1}{4}$ inch | 0.25 |
| $\frac{5}{16}$ inch | 0.31 |
| $\frac{3}{8}$ inch | 0.38 |
| $\frac{7}{16}$ inch | 0.44 |
| $\frac{1}{2}$ inch | 0.50 |
| $\frac{9}{16}$ inch | 0.56 |
| $\frac{5}{8}$ inch | 0.63 |
| $\frac{11}{16}$ inch | 0.69 |
| $\frac{3}{4}$ inch | 0.75 |
| $\frac{13}{16}$ inch | 0.81 |
| $\frac{7}{8}$ inch | 0.88 |
| $\frac{15}{16}$ inch | 0.94 |

**Table 22-8** Boiling Points of Water at Various Pressures Above Atmospheric

| Gage Pressure (psi or kPa) | Boiling Point |
|---|---|
| 1–6.89 | 216°F (102.2°C) |
| 4–27.58 | 225°F (107.2°C) |
| 15–103.43 | 250°F (121.1°C) |
| 25–172.36 | 267°F (130.5°C) |
| 30–206.84 | 274°F (134.4°C) |
| 45–310.26 | 293°F (145.0°C) |
| 50–344.73 | 297°F (147.2°C) |
| 65–448.13 | 312°F (155.5°C) |
| 75–517.1 | 320°F (160.0°C) |
| 90–620.52 | 335°F (168.3°C) |
| 100–689.47 | 338°F (170.0°C) |
| 125–861.83 | 353°F (178.3°C) |
| 150–1034.2 | 366°F (185.5°C) |

Atmospheric pressure of 101.3 kPa will balance or support a column of mercury 76 cm high.

**Table 22-9** Weight (Mass)

| Measurement | Equivalent |
|---|---|
| 1 kg | 2,204623 pounds |
| 453.592 g | 1 kg |
| 1 g | 0.035 ounce |
| 28.349 g | 1 ounce |
| 28,349 mg | 1 ounce |
| 1 g | 1,000 mg |
| 1 kg | 1,000,000 mg |
| 1 kg | 1,000 g |
| 0.02831 kg | 1 ounce |
| 1 lb | 453,592.37 mg |
| 1 lb | 453.59837 g |
| 1 lb | 0.453592 kg |
| 1 metric ton | 1,000 kg |
| 1 metric ton | 2,204.623 pounds |
| 1 mL water | 1 g |
| 1 L cold water (40°C) | 1 kg |

# Review Questions

1. What is the basic unit of measurement in the metric system of measurement?

   _____

2. What does the unit mole measure?

   _____

3. What is the lumen used to measure?

   _____

4. What is the basic unit of time?

   _____

5. At what temperature does water boil at sea level?

   _____

6. How many degrees are there in a circle?

   _____

7. How many 90-degree angles will make a circle?

   _____

8. What is the metric system's basic unit for length?

   _____

9. What is the value of the Greek letter pi?

   _____

10. How many sides does a triangle have?

    _____

# 23 | Poisons and Hazardous Materials

## Performance Objectives

After studying this chapter, you will:

- Understand what poisons a plumber is exposed to.

- Know which household products to be aware of.

- Know who to call for accidental poisoning.

- Know which acid reacts with which pipe type.

- Know how and when to work with dangerous materials.

- Know how to recognize poisoned people.

- Know who to contact for proper clothing for a particular job.

- Know where and how to dispose of poisons encountered on the job.

- Be able to answer the review questions at the end of this chapter.

Most state plumbing codes and licensing examinations are designed to establish environmental sanitation and safety through properly designed supervision that will ensure properly installed and maintained plumbing systems. Details of plumbing construction vary, but the basic sanitary and safety principles are the same. The desired and required results (to protect the health of people) are similar regardless of locality. These basic principles require that all plumbing in public and private buildings intended for human occupation or use be installed so that it is capable of protecting the health, welfare, and safety of the occupants and the public. Deserving of attention are the poisons in household products and hazardous materials that have a habit of cropping up in plumbing systems—both in residences, stores, industries, and schools. The plumber must be very aware of these materials.

## Basic Plumbing Principles

Basic plumbing principles include the following:

- Buildings intended for human occupancy or use will be provided with a supply of pure and wholesome water with connections not subjected to the hazards of backflow or back siphonage and not connected to unsafe water supplies. If there is a public water main available, an individual connection to the public water main shall be made.
- Buildings with plumbing fixtures and devices shall be provided with a supply of water in sufficient volume and pressure to enable them to operate in a satisfactory manner at all times.
- Water heaters and other devices used for purposes of water heating and storing shall be designed and installed to prevent explosion through overheating.
- Where public sanitary sewers are available, buildings intended for human occupancy shall have a connection made to the public sanitary sewer.
- Plumbing fixtures shall be made of materials that are durable, corrosion-resistant, non-absorbent, and free of concealed fouling surfaces. Rooms in which water closets, urinals, and similar fixtures are installed shall have proper ventilation and adequate lighting.
- It is recommended that family dwelling units adjacent to sanitary sewer lines or having private sewage disposal systems have at least one lavatory, one water closet, a bathtub or shower, and a kitchen-type sink for purposes of sanitation and personal hygiene. Other structures for human occupancy with sanitary public or private sewage disposal systems should have no less than one water closet and one fixture for hand-washing.
- Building sanitary drainage systems shall be designed, installed, and maintained in a condition so as to conduct wastewater and sewage to designated locations from each fixture with a flow that prevents fouling, clogging, and deposits of solids in the piping. Sufficient cleanout shall be installed so that the piping system can be easily cleaned in case of stoppage.
- Plumbing systems shall be maintained in a sanitary condition, and each connection (direct or indirect) to the drainage system shall have a water-seal trap. The system shall be kept in a serviceable condition with adequate spacing of the fixtures. These fixtures should be reasonably accessible for cleaning.
- Drainage pipe shall be designed and installed with a durable material free of water leakage and offensive odors caused by drain sewer air. Installation shall be in accordance with good workmanship practices and use of good grade material.

- Plumbing systems shall be designed, installed, and kept in adjustment so as to provide the required quantity of water consistent with adequate performance. There should be no undue noise under normal conditions and use. New systems and/or remodeled systems shall be subjected to tests that will disclose leaks and defects.

- Included in the design shall be every consideration for the preservation of the strength of structural members of the building. Each vent terminal extending to the outer air shall be designed to minimize clogging and the return of foul air to the building.

- Design considerations shall include protection from contamination by sewage backflow of water, food, disposal of sterilized items, and similar materials.

- Substances that will clog pipes or their joints and interfere with the sewage disposal process or produce explosive mixtures shall not be allowed in the building sewage drainage system.

- Sewage or other wastes from a plumbing system shall not discharge into subsurface soil or into a water surface unless it has first been treated in an acceptable manner.

## Acids

Acids are chemicals that turn litmus paper red. Litmus is a colored paper strip that can change from red to blue and back again. Its color depends upon the concentration of hydrogen ions being tested. If the concentration of hydrogen ions is higher than in pure water, the litmus turns red. If it is lower than in pure water, the litmus turns blue.

pH is a measure of the concentration of hydrogen ions. Hydrogen chloride is a gas; it consists of molecules each containing one atom of hydrogen and one atom of chlorine. When this gas is dissolved in water, the molecules of hydrogen chloride disassociate. The hydrogen and chlorine atoms become separated. The chlorine atom obtains an extra electron and becomes a chlorine ion. The hydrogen atom loses its electron and becomes a hydrogen ion. Hydrogen ions have a positive charge and chloride ions have a negative charge. When hydrogen ions combine with litmus, the litmus turns red. When litmus loses its hydrogen ions it turns blue. An alkali is the opposite of an acid. Alkalis produce virtually no hydrogen ions.

## Strong Acids

- **Sulfuric Acid ($H_2SO_3$).** One of the strong acids. It can eat skin, cloth, and almost anything it comes in contact with. It is found in many manufacturing practices and products. It is often encountered in the lead-acid batteries used in all cars and trucks. When mixed with water a violet reaction results. Always mix the acid in a large quantity of water. Do it slowly to control the reaction. Beware of splattering. Protect your eyes and skin at all times when working around sulfuric acid.

- **Hydrochloric Acid (HCl).** Another of the strong acids, it is found in the human stomach to aid in the digestive process. Hydrochloric acid is also used in the mining and refining of gold. The gold combines with the chlorine atom and produces gold chloride, which is then processed further to release the chlorine, leaving only the gold. The chlorine is a gaseous form. It was used in World War I (1914–1918) as a poisonous gas by the Germans.

- **Nitric Acid ($HNO_3$).** Another of the inorganic (strong) acids with many industrial uses. It is used in producing metals, plastics, explosives, textiles and various dyes.

Many acids are deadly poisons, even though they are made up of harmless common elements such as carbon, oxygen, and hydrogen. These poisons include oxalic acid and carbolic acid (phenol). Some of the acids encountered every day are:

- Acetic acid
- Amino acid
- Aqua regia boric acid carbolic acid
- Citric acid formic acid
- Gallic acid
- Hydrochloric acid
- Hydrofluoric acid
- Lactic acid
- Nitric acid oxalic acid
- Phosphoric acid
- Picric acid

- Prussic acid
- Salicylic acid (used in aspirin and other common medicines)
- Stearic acid
- Sulfuric acid tannie
- Acid tartaric acid

Plumbers need to know about acids and how they produce corrosion of pipes and fixtures and the results of their actions with metals and plastics. Acids can affect the health of those who work with or around them (Tables 23-1 to 23-3).

**Table 23-1**    Acid Resist Chart for PVC Pipe and Fittings (Femco, Inc.)

| Reagent | PVC 72° | Reagent | PVC 72° | Reagent | PVC 72° |
|---|---|---|---|---|---|
| Acetic Acid 20% | R | Cottonseed Oil | N | Plating Solution | R |
| Acetic Acid 80% | N | Cresol | N | Potassium Carbonate | R |
| Acetone | N | Cyclohexanol | N | Potassium Chlorate | R |
| Alcohol (Methyl or Ethyl) | R | Cyclohexanone | N | Potassium Chloride | R |
| Aluminum Chloride | R | Dimethylamine | N | Potassium Cyanide | R |
| Aluminum Sulfate | R | Dioctyl Pthalate | N | Potassium Dichromate | R |
| Alums | R | Disodium Phosphate | N | Potassium Hydroxide | R |
| Ammonia Gas (Dry) | R | Distilled Water | R | Potassium Permanganate 10% | R |
| Ammonium Chloride | R | Ethers | N | Potassium Sulfate | R |
| Ammonium Hydroxide | R | Ethyl Acetate | N | Propane Gas | R |
| Ammonium Nitrate | R | Ethylene Chloride | N | Propyl Alcohol | R |
| Ammonium Phosphate | R | Ethylene Glycol | N | Sea Water | R |
| Ammonium Sulfate | R | Fatty Acids (C6) | R | Sewage | R |
| Ammonium Sulfide | R | Ferric Chloride | R | Silver Cyanide | R |
| Amyl Chloride | N | Ferric Sulfate | R | Silver Nitrate | R |
| Aniline | N | Fluorine (Gas Wet) | N | Silver Sulfate | R |
| Aqua Regia | N | Formaldehyde (20%) | N | Sodium Bicarbonate | R |
| Barium Chloride | R | Formic Acid (10%) | N | Sodium Bisulfite | R |
| Barium Hydroxide | R | Freon 12 Dry | N | Sodium Carbonate | R |
| Barium Sulfate | R | Fruit Juices & Pulp | R | Sodium Cyanide | R |
| Barium Sulfide | R | Furfural | N | Sodium Ferrocyanide | R |
| Beer | R | Gasoline (Refined) | N | Sodium Hydroxide | R |
| Beet Sugar Liquors | R | Glucose | R | Sodium Hypochlorite | R |
| Benzene | N | Glycerine | N | Sodium Sulfate | R |
| Benzoic Acid | R | Hydrobromic Acid (20%) | N | Sodium Sulfide | R |
| Black Liquor | N | Hydrochloric Acid | R | Sodium Sulfite | R |
| Bleach 12 5% active Cl2 | R | Hydocyanic Acid | N | Sodium Thiosulfate | R |
| Boric Acid | R | Hydoquinone | R | Stannic Chloride | R |
| Bromic Acid | R | Hypochlorous Acid | R | Stannous Chloride | R |
| Bromine Water | N | Iodine | N | Stearic Acid | R |
| Butane | N | Kerosene | N | Sulfite Liquors | R |
| Butyric Acid | N | Lactic Acid 25% | R | Sulfur | R |
| Calcium Carbonate | R | Linseed Oil | N | Sulfur Dioxide (Dry) | R |
| Calcium Chloride | R | Liquors | N | Sulfur Dioxide (Wet) | R |
| Calcium Hydroxide | R | Machine Oil | N | Sulfuric Acid 50% | R |
| Calcium Hypochlorite | R | Magnesium Chloride | R | Sulfuric Acid 70% | R |
| Calcium Sulfate | R | Magnesium Sulfate | R | Sulfuric Acid 93% | N |
| Cane Sugar Liquors | R | Maleic Acid | N | Sulfurous Acid | N |
| Carbon Bisulfide | N | Methyl Chloride | N | Tannic Acid | R |
| Carbon Dioxide | R | Methyl Ethyl Ketone | N | Tanning Liquors | R |
| Carbon Monoxide | R | Milk | R | Tartaric Acid | R |
| Carbon Tetrachloride | N | Mineral Oils | N | Toluene | N |
| Carbon Acid | R | Mixed Acids | R | Trichloroethylene | N |
| Caustic Soda | R | Muriatic Acid | R | Triethanolamine | N |
| Caustic Soda 50% | R | Nickel Chloride | R | Trisodium Phosphate | N |
| Caustic Potash | N | Nickel Sulfate | R | Turpentine | N |
| Chloride (Dry) | N | Oils & Fats | N | Urea | R |
| Chloride (Wet) | N | Oleic Acid | N | Urine | R |
| Chloroacetic Acid | N | Oelum | N | Vinegar | R |
| Chlorobenzene | N | Oxalic Acid | R | Water (Fresh) | R |
| Chloroform | N | Palmitic Acid 10% | N | Water (Salt) | R |
| Chromic Acid 10% | N | Perchloric Acid 10% | R | Whiskey | R |
| Chromic Acid 50% | N | Perchloric Acid 70% | N | Wines | R |
| Citric Acid | R | Petroleum Oils (Sour) | N | Xylene | N |
| Copper Chloride | R | Phenol 5% | N | Zinc Chloride | R |
| Copper Cyanide | R | Photographic Solutions | R | Zinc Sulfate | R |
| Copper Nitrate | R | Photographic Solutions | N | | |
| Copper Sulfate | R | Picric Acid | N | | |

R = Recommended; N = Not Recommended
Note: The data listed in this table is only to give information in regard to general use and does NOT constitute a guarantee.
Materials should be tested under actual service to determine suitablility for a particular purpose.

**Table 23-2**   Common Poisons for Children and Adults (National Capital Poison Center)

## The Most Common Poisons in Children
Did you know that even these common houshold items can poison children?
- Cosmetics and personal care products
- Cleaning substances and laundry products
- Foreign bodies such as toys, coins, thermometers
- Topical preparations
- Vitamins
- Antihistamines
- Pesticides
- Plants
- Antimicrobials

## The Most Common Poisons in Adults
- Pain medicine
- Sedatives, hypnotics, antipsychotics
- Antidepressants
- Cardiovascular drugs
- Cleaning substances
- Alcohols
- Pesticides
- Bites and envenomations by ticks, spiders, bees and snakes
- Anticonvulsants
- Cosmetics and personal care products

## The Most Dangerous Poisons for Children
These are especially hazardous household items. Buy small quantities. Discard unneeded extras. Make sure they are always out of a child's reach.
- Medicines
- Carbon monoxide
- Button batteries
- Iron pills
- Cleaning products
- Nail glue remover or nail primer
- Hydrocarbons: gasoline, kerosene, motor oil, lighter fluid, furniture polish, and point thinner
- Pesticides
- Windshield washer solution and antifreeze
- Wild mushrooms
- Alcohol
- Drain cleaners and toilet bowl cleaners
- Topical anesthetics

**Table 23-3** Common Drinking Water Contaminants (Neilson Research Corp and the EPA)

All sources of drinking water contain some naturally occurring contaminants. As water flows in streams, sits in lakes, and filters through layers of soil and rock, it dissolves or absorbs the substances that it touches. According to its exposure, water transforms in composition and in physical parameters.

| | |
|---|---|
| Aluminum (Al) | Low level exposure is not thought to harm your health. Aluminum, however, is not a necessary substance for our bodies and too much may be harmful. (Federal Limit 0.05 – 0.2 mg/L) |
| Antimony (Sb) | Above the EPA limit antimony may potentially cause nausea, vomiting, and diarrhea. Antimony is a known/potential drinking water human carcinogen. (Federal Limit 0.006 mg/L) |
| Arsenic (As) | Arsenic is a known human carcinogen. (Federal Limit 0.010 mg/L) |
| Barium (Ba) | Symptoms of barium poisoning include increased blood pressure, changes in heart rhythm, stomach irritation, and muscle weakness. (Federal Limit 2.0 mg/L) |
| Beryllium (Be) | Beryllium is a probable human carcinogen. (Federal Limit 0.004mg/L) |
| Boron (B) | Exceptionally toxic to some plants. If you have problems with growing plants, it could be the water and not your green thumb! (Toxic range for plants is 1.0-4.0 mg/L) |
| Cadmium (Cd) | Symptoms of cadmium poisoning include cramps, nausea, vomiting, and diarrhea. Long term exposure to lower levels of cadmium leads to kidney disease, lung damage and fragile bones. (Federal Limit 0.005mg/L) |
| Calcium (Ca) | Calcium is an important contributor to water hardness. (No Federal Limit) |
| Conductivity | Conductivity gives an approximate determination of the amount of dissolved minerals in the water. (No Limit) |
| Chromium (Cr) | Above the EPA limit chromium may potentially cause skin irritation or ulceration. Long term exposures to chromium may cause damage to liver, kidney, circulatory, and nerve tissues. (Federal Limit 0.1 mg/L) |
| Copper (Cu) | Causes staining of fixtures, hair, and fabrics and can impart a bitter taste to water. It can cause stomach irritation and vomiting. (Federal Limit 1.0 mg/L) |
| Fecal Coliform Bacteria and Escherichia coli (E. coli) | Present in the intestines of mammals. In the laboratory, coliforms are used as indicators of fecal contamination of ground and surface waters. Water sources containing any coliforms must be treated before consumption. |
| Fluoride (F) | Long term effects are a permanent brown staining of the teeth, destruction of tooth enamel, brittle and easily broken bones, painful and stiff joints. (Federal Limit 4.0 mg/L, Oregon limit 2.0 mg/L) |
| Hardness | Hardness is usually attributed to the calcium and magnesium ions. These ions combine with soap, forming an insoluble precipitate visible as scum and rings around fixtures. (Federal Limit 250 mg/L) |
| Iron (Fe) | When iron comes in contact with oxygen, it oxidizes to a visible reddish compound that settles out as a rust-like material that stains clothing and fixtures. (Federal Limit 0.3 mg/L) |
| Lead (Pb) | Symptoms of lead poisoning include tiredness and aching bones. (Federal Limit 0.015 mg/L) |
| Lithium (Li) | Occurs naturally in Southern Oregon and is currently being monitored by NRC. (No Limit) |
| Magnesium (Mg) | Magnesium is an important contributor to water hardness. When water is heated, magnesium breaks down and precipitates out of solution, forming scale. Magnesium concentrations greater than 125 mg/L may have a laxative effect. (No Limit) |
| Manganese (Mn) | Produces a brownish discoloration, which stains clothing and fixtures. High levels of manganese are toxic to expectant mothers and children. (Federal Limit 0.05 mg/L) |
| Molybdenum | Excessive molybdenum consumption can be associated with enlarged liver, gastrointestinal, and kidney disorders. (USEPA Lifetime Health Advisory: 40 ug/L) |
| Nickel (Ni) | Relatively short exposures above the EPA Limit are not known to cause any health problems. Long term exposures can potentially cause decreased body weight, skin irritation, heart, and liver damage. (Federal Limit 0. 1 mg/L) |
| Nitrate/Nitrite ($NO_2$/$NO_3$) | Affects infants under the age of 6 months. In this age group nitrates reduce the blood's ability to carry oxygen and may cause death or permanent brain damage. (Federal Limit Nitrate 10 mg/L, Nitrite 1 mg/L) |
| Pesticides and Herbicides | Enter surface and ground water primarily as runoff and can remain in sediment for years. Thousands of chemicals are currently regulated by the EPA and have various hazardous effects on humans. (Federal Limit Per Each Analyte) |

**Table 23-3**   Common Drinking Water Contaminants (Neilson Research Corp and the EPA) (*continued*)

| | |
|---|---|
| pH | The ideal pH for drinking water is 7.5. When pH is below 7.0, the water is acidic and can cause corrosion of pipes and fixtures. When the pH is higher than 8.0, the water is alkaline. This can create mineral deposits on the interior surfaces of pipes. |
| Potassium (K) | To lower blood pressure, blunt the effects of salt, and reduce the risk of kidney stones and bone loss, adults should consume 4.7 grams of potassium per day. (No Limit) |
| Selenium (Se) | Is an essential nutrient at low levels. However, levels above 0.05 ppm may cause: hair and fingernail changes; damage to the peripheral nervous system; fatigue and irritability. Long-term exposures to selenium may cause hair and fingernail loss, damage to kidney and liver tissue and the nervous and circulatory systems. (Federal Limit 0.05 mg/L) |
| Silica | Silica analysis provides useful information for systems that my require water treatment. Not identified as a health hazard. (No Limit) |
| Silver (Ag) | Silver poisoning causes a blue-gray discoloration of the skin, mucous membranes, and eyes. In high doses it is fatal to humans. (Federal Limit 0.1 mg/L) |
| Sodium and Chloride (Na/Cl2) | If the sodium and chloride levels are near 100 mg/L, individuals may notice a salty taste. These levels also affect plant growth. (Sodium: No Limit) (Chloride: Federal Limit is 250 mg/L) |
| Sulfate (SO$_4$) | Sulfate is a substance that occurs naturally. It may be found in the form of hydrogen sulfide and is commonly identified by a "rotten egg odor." Diarrhea may be associated with the ingestion of high levels of sulfate. (Federal Limit 250 mg/L) |
| Thallium (Tl) | Above the EPA limit, thallium may potentially cause gastrointestinal irritation and nerve damage. Long-term exposures to thallium may cause changes in blood chemistry, hair loss, damage to liver, kidney, intestinal, and testicular tissues. (Federal Limit 0.002 mg/L) |
| Turbidity | Turbidity is the lack of clarity or brilliance in water. This can affect water treatment systems such as UV lights for disinfection, reverse osmosis units, sediment removal systems, and ion exchange treatment systems. (Federal Limit 1 NTU) |
| Uranium | Naturally occurring substance that is mildly radioactive. Exposure to high levels of uranium can cause kidney disease. (Federal Limit 0.03 mg/L) |
| Vanadium | The health effects in humans has not been established. Studies in pregnant animals showed minor birth defects. Vanadium ingested over a long period of time also revealed minor kidney and liver changes. Vanadium is also used for arsenic removal in drinking water treatment systems. (No Limit) |
| Volatile Organics (VOCs) | VOCs are found in gasoline, dry cleaning solvents, degreasing agents and other industrial solutions. The EPA and DEQ monitor thousands of chemicals that fall under this classification. (Federal Limit Per Each Analyte) |
| Zinc (Zn) | High levels of zinc can cause stomach cramps, nausea, and vomiting. Over a long period of time, it can cause anemia and pancreas damage. (Federal Limit 5.0 mg/L) |

# Review Questions

1. What acid is found in automobile batteries?

   _____

2. What is the formula for sulfuric acid?

   _____

3. Where would you expect to find acetic acid?

   _____

4. What common acid is found in aspirin?

   _____

5. What industrial process uses hydrochloric acid?

   _____

6. Where does the plumber find dangerous bowl cleaning acid?

   _____

7. Is alkali another word for acid?

   _____

8. Are acids organic or inorganic?

   _____

9. What does pH stand for?

   _____

10. How is the pH of water measured?

    _____

# 24 | Hand Signals for Hoisting and Knots

## Performance Objective

After studying this chapter, you will:

- Be able to identify a square knot.

- Understand the role knots and hand signals play in plumbing supplies handling.

- Be able to tie the various knots shown in Figure 24-1.

- Be able to perform the signals shown in Figure 24-2.

- Be able to answer the review questions at the end of this chapter.

Hand signals are another way of communicating. There are standard signals utilized for those who work in a noisy area or cannot be heard by other means.

Boy Scouts spend a lot of time learning their knots and so do those who work with the kinds of equipment that hoist heavy loads and such things as stacks of pipe.

Large concrete pipes and heavy steel pipes may need to be moved to another location and in order to do so efficiently and safely, a good knot in the rope or cable will work well. For the knots that are most often used and are commonly understood by plumbers, see Figure 24-1. These knots are usually appropriate for most of their work requirements.

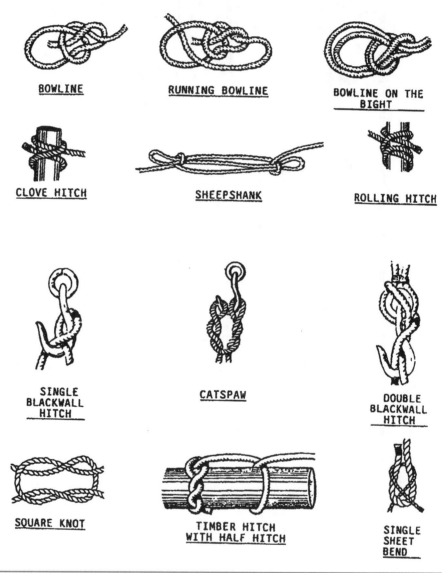

**Figure 24-1**   Common knots.

**Figure 24-2**   Hand signals.

## Review Questions

**1.** Where are hand signals used?

_____

**2.** Why would you, as a plumber, need hand signals?

_____

**3.** List at least three types of knots.

_____

**4.** Describe the square knot.

_____

**5.** Where do plumbers use knots?

_____

# 25 License Requirements and Applications

## Performance Objectives

After studying this chapter, you will:

- Learn the license requirements for your state.

- Know the fees associated with special licenses.

- Know who to contact for information on licensing of plumbers, journeyman plumbers, master plumbers, and contractors.

- Know the purpose of licensing those who work in this trade.

- Be able to answer the review questions at the end of this chapter.

Individual states have their own requirements for obtaining a license to be a *journeyman plumber*, a *plumbing contractor*, a *master plumber*, or an *apprentice*. Some states break licensing into specialty areas such as irrigation contractor, irrigation installer, appliance installer, mobile home contractor, mobile home installer, sewer and water installer, sewer and water contractor, water conditioning contractor, and water conditioning installer. The schedule of fees for each state varies so much that the amounts printed today may be changed tomorrow. Some states and local communities see these licensing fees as another way to support the plumbing department (both inspectors and administrators) without tapping into local tax revenues. Then again, other communities will accept the state plumber's license as proof of your qualifications (Table 25-1).

**Table 25-1**   License Application Requirements, by State

LICENSE APPLICATION REQUIREMENTS

Apprentice, journeyman, contractor, and master plumber license application requirements vary from state to state. For information about the application process, visit your state's Web site.

| State | Organization | Official Web Site |
|---|---|---|
| Alabama | Plumbers and Gas Fitters Examining Board | www.pgfb.state.al.us/ |
| Alaska | Department of Community and Economic Development | www.dced.state.ak.us/occ/ |
| Arizona | Registrar of Contractors | www.azroc.gov/ |
| Arkansas | Contractor's Licensing Board | www.state.ar.us/clb/ |
| California | Contractor's State License Board | www.cslb.ca.gov/ |
| Colorado | Examining Board of Plumbers | www.dora.state.co.us/Plumbing/ |
| Connecticut | Department of Consumer Protection | www.state.ct.us/dcp |
| Delaware | Board of Plumbing, Heating, Ventilation, Air Conditioning, and Refrigeration Examiners | http://dpr.delaware.gov/boards/plumbers/index.shtml |
| Florida | Department of Business and Professional Regulation | www.myflorida.com/dbpr/ |
| Georgia | State Construction Industry Licensing Board | www.sos.state.ga.us/plb/ |
| Hawaii | Department of Commerce and Consumer Affairs | www.state.hi.us/dcca |
| Idaho | Plumbing Bureau | www2.state.id.us/dbs/plumbing/index.html |
| Illinois | Department of Professional Regulation | www.idfpr.com/dpr/default.asp |
| Indiana | Professional Licensing Agency | www.in.gov/pla/plumbing.htm |
| Iowa | Plumbing and Mechanical Systems Board | www.idph.state.ia.us |
| Kansas | Department of Revenue | www.ink.org/public/kdor/ |
| Kentucky | Department of Housing, Buildings, and Construction | http://ohbc.ky.gov/plb/plblic.htm |
| Louisiana | State Plumbing Board | www.legis.state.la.us |
| Maine | Plumbers' Examining Board | www.state.me.us/pfr/olr |
| Maryland | Board of Plumbing | www.dllr.state.md.us/license/plumb/plumbintro.html |
| Massachusetts | Division of Professional Licensure | www.state.ma.us/reg/boards/pl/default.htm |
| Michigan | Department of Labor | www.michigan.gov/cis/ |
| Minnesota | Department of Labor and Industry | www.dli.mn.gov/ccld/plumbing.asp |
| Mississippi | Contractor Licensing Board | www.msboc.us/ |
| Missouri | Department of Labor and Industrial Relations | www.dolir.mo.gov/ |
| Montana | State Board of Plumbers | http://mt.gov/dli/bsd/license/bsd_boards/plu_board/board_page.asp |
| Nebraska | Workforce Development—Department of Labor | www.dol.state.ne.us/ |
| Nevada | State Contractors Board | www.nscb.state.nv.us |

**Table 25-1**    License Application Requirements, by State (*continued*)

| State | Organization | Official Web Site |
|-------|--------------|-------------------|
| New Hampshire | State Board for Licensing and Regulating Plumbers | www.state.nh.us/plumbing/ |
| New Jersey | Department of Law and Publc Safety | www.state.nj.us/lps/ca/nonmedical/plumbers.htm |
| New Mexico | Construction Industries Division | www.rld.state.nm.us/CID/index.htm |
| New York | Department of Buildings | http://www1.nyc.gov/site/buildings/industry/licensing.page |
| North Carolina | State Board of Examiners of Plumbing, Heating, and Fire Sprinkler Contractors | www.nclicensing.org |
| North Dakota | State Plumbing Board | www.governor.state.nd.us/boards/boards-query.asp?Board_ID=83 |
| Ohio | Department of Commerce | www.com.state.oh.us/ |
| Oklahoma | State Department of Health | www.cib.state.ok.us/ |
| Oregon | Consumer & Business Services Building Codes Division | www.oregonbcd.org |
| Pennsylvania | Association of Plumbing, Heating, and Cooling Contractors | www.paphcc.org/ |
| Rhode Island | Department of Labor and Training | www.dlt.state.ri.us/profregs/ |
| South Carolina | Contractor's Licensing Board | www.llr.state.sc.us/POL/ResidentialBuilders/ |
| South Dakota | State Plumbing Commission | http://dol.sd.gov/bdcomm/plumbing/pclicensereqs.aspx |
| Tennessee | Board for Licensing Contractors | http://tn.gov/commerce/boards/contractors/index.shtml |
| Texas | State Board of Plumbing Examiners | www.tsbpe.state.tx.us/ |
| Utah | Division of Occupational and Professional Licensing | www.commerce.state.ut.us |
| Vermont | Plumbers Licensing Board | http://firesafety.vermont.gov/professions/plumbing |
| Virginia | Department of Professional and Occupational Regulation | www.dpor.virginia.gov/dporweb/dpormainwelcome.cfm |
| Washington | Department of Labor and Industries | www.lni.wa.gov/TradesLicensing/Contractors/HowReg/default.asp |
| Washington, D.C. | Department of Consumer & Regulatory Affairs | http://dcra.dc.gov/dcra/site/default.asp |
| West Virginia | Contractor Licensing Board | www.labor.state.wv.us |
| Wisconsin | Department of Commerce, Safety and Buildings Division | http://commerce.wi.gov/ |
| Wyoming | Department of Employment | http://doe.wyo.gov |

Data from Contractor's License Reference Site (www.contractors-license.org)

An apprentice should have the required background knowledge and have the required documents for certifying your experience and schooling properly notarized at the time you make your application for the exam. The best way to obtain information is on the Internet. Each state has a website that details the requirements and lists the fees attached to each specialty area. Usually you can contact the proper office by www.*<name of the state>*.us.org.

In this chapter we provide a sampling of states such as Alabama, Colorado, Delaware, Indiana, Michigan, Nevada, New Jersey, New Mexico, South Dakota, Texas, and Utah.

Some cities, such as Las Vegas, Nevada, require a plumbing license for the county/city.

New York City, of course, has its own licensing bureau with requirements and tests.

No matter where the test is taken, certain facts, figures, and techniques must be mastered before the state will grant a license and certify that the applicant is qualified to do the work covered by the license. This procedure is necessary to ensure that the water and sewage systems of the nation function properly and with maximum efficiency. A few selected states have been chosen here to illustrate the process employed by each state to ensure clean water.

## Alabama

Table 25-2 shows the requirements and fees for certification in the state of Alabama. Keep in mind that the fees stated here may not be correct; up-to-date fee schedules are available at your local licensing facility. The fees here are for the purpose of illustration of the amount for the various licenses. These fees change each year as the need arises, so the latest update on fees must be obtained by using the Internet.

## Colorado

Colorado application requirements for journeyman plumber and master plumber can be found at https://www.colorado.gov/pacific/dora/ Plumbing_Licensing_Requirements. Note that the journeyman fee is $67 and the master plumber fee is $107. The fees are subject to change every July 1. Colorado has a statewide plumbing license for residential, journeymen, and masters.

## Delaware

The State of Delaware Board of Plumbing Examiners has statutory authority to grant plumbers licenses. There are five pages concerning the licensing of plumbers on the Internet (https://dpr.delaware.gov/boards/plumbers/newlicense/). The rules and regulations cover everything from general provisions to the voluntary treatment option for chemically dependent or impaired professionals.

**Table 25-2**   Requirements and Fees for Certification in Alabama

| Requirements | Renewal Fees | Non-Refundable Exam Fees |
|---|---|---|
| Apprentice | $25.00 (Initial w/ Registration) | $25.00 (Annually) |
| No written exam required. Complete registration form and include a copy of proof of citizenship or lawful presence of non-citizen form. | | |
| Journeyman Plumber | $50.00 | $110.00 |
| Requires 2 years current work experience as Apprentice or completed Board Approved Apprentice Training Program to take required exam. | | |
| Journeyman Gas Fitter | $50.00 | $110.00 |
| Requires 2 years current work experience as Apprentice or completed Board Approved Apprentice Training Program to take required exam. | | |
| Master Plumber | $150.00 | $160.00 |
| Requires one year current work experience as Journeyman Plumber to take required exam. | | |
| Master Gas Fitter | $150.00 | $160.00 |
| Requires one year current work experience as Journeyman Gas Fitter to take required exam. | | |
| Medical Gas | $25.00 | |
| Requires certification from National Institute of Testing & Certification. | | |
| Replacement Card | $25.00 | |

## Indiana

The State of Indiana has its application forms and answers to frequently asked questions (FAQs) posted on the Internet (www.in.gov/pla/bandc/plumbing/plumbing faq.html). Following are some of the questions and answers:

**Question:** How do you become an apprentice?

**Answer:** You must be registered in an approved program and be 17 years or older to qualify as an apprentice.

**Question:** How do you become a journeyman?

**Answer:** You must have at least four years in an apprenticeship program, or four years of experience in the plumbing field to qualify as a journeyman.

**Question:** How do you become a contractor?

**Answer:** You must have four years in an apprenticeship program or four years of experience in the plumbing trade or plumbing business under the direction of a licensed plumbing contractor.

**Question:** How many times can you take the test?

**Answer:** You can take the test seven times within a two-year period. However, if you do not pass the entire exam on your first attempt, you shall be entitled to take the exam six additional times.

**Question:** Does reciprocity exist for journeyman and contractor plumbers?

**Answer:** Reciprocity is not available for journeyman and contractor plumbers. Instead, you must present an application and pay the appropriate fee. Then the application goes before the board. If the board approves, you must take the exam and pass.

**Question:** How much is the apprentice fee?

**Answer:** The apprentice fee is $10.

**Question:** How much is the journeyman fee?

**Answer:** The journeyman application fee is $15.

**Question:** How much is the application fee for contractor?

**Answer:** The application fee for contractor is. $30.

**Question:** May I work as a plumber without a license?

**Answer:** It is unlawful to act in the capacity of a plumbing contractor or journeyman plumber without a license.

The application form for approval of plumbing apprentice school displayed in Figure 25-1 shows the extent to which the state goes to make sure the apprenticeship programs are run on a professional level. The Indiana Professional Licensing Agency is responsible for making sure the schools meet the state laws and rules of the Indiana Plumbing Commission. Keep in mind that the fee amounts are subject to change. They are given here so you can get an idea of the difference between the various plumbing fees. As an example, notice that in the application for journeyman, the plumber examination fee is $30. The fees are non-refundable and non-transferable in case you don't show up for the test.

## Michigan

As in any state, a license is required to be a plumbing contractor. To qualify for the license, the applicant must hold a master plumber license or be in the employ of a holder of a master plumber license as his or her representative. Only an owner of a sole proprietorship or partnership, or an officer of a corporation or limited liability company, may apply for licensure as a plumbing contractor.

The application for plumbing contractor shown in Figure 25-2 provides a glimpse of the requirements in this particular state. Other states are similar. In this case, the application is issued by the Michigan Department of Consumer & Industry Services, Bureau of Construction Codes & Fire Safety, and Plumbing Division in the state capital (Lansing).

Apprentices must serve a four-year apprenticeship or, in some instances (for specialized areas), a two-year apprenticeship is required. Documentation of experience is required by each state to become a journeyman, a master plumber, or a plumbing contractor. This is in addition to taking and passing the required exam for the area of specialization. Each exam has a healthy fee attached to it. All exams can be retaken, but this policy varies from state to state. There is a retake fee if you fail the first time.

**Figure 25-1** Indiana application form

## APPLICATION FOR JOURNEYMAN PLUMBER EXAMINATION FOR LICENSING
State Form 40602 (R19 / 8-16)
Approved by State Board of Accounts, 2016

**INDIANA PLUMBING COMMISSION**
**PROFESSIONAL LICENSING AGENCY**
402 West Washington Street, Room W072
Indianapolis, Indiana 46204-2724
Telephone: (317) 234-8800
E-mail: pla14@pla.IN.gov
www.pla.IN.gov

INSTRUCTIONS:   1.  Please refer to the instructions on our website, www.pla.in.gov, for the licensing requirements and http://www.in.gov/pla/2762.htm for the fee in accordance with 860 IAC 1-1-2.1.
2.  All fees are non-refundable and non-transferable.

\* This agency is requesting the disclosure of your Social Security number in accordance with Indiana Code 4-1-8-1. Disclosure is mandatory; this record cannot be processed without it. Social Security numbers will be made available to the Department of Revenue.

### FOR OFFICE USE ONLY

| Application fee | Date fee paid (month, day, year) | Receipt number |
|---|---|---|
| License number | | Date issued (month, day, year) |

### DO NOT WRITE ABOVE THIS LINE

### SECTION 1 - APPLICANT INFORMATION (To be completed by all applicants.)

| Name (last, first, middle, maiden) | Social Security number * |
|---|---|

Address (number and street)

City, state, and ZIP code

| Date of birth (month, day, year) | Place of birth |
|---|---|

| Telephone number ( ) | E-mail address |
|---|---|

Are you the spouse of a member of the military who is assigned to a duty station in Indiana (Optional)?   ☐ Yes   ☐ No

List all states in which you hold or have held a plumbing related professional or trade license, registration, or permit.

| TYPE OF LICENSE | STATE | NUMBER | DATE OF ISSUE (month, day, year) | STATUS |
|---|---|---|---|---|
| | | | | |
| | | | | |

### PERSONAL BACKGROUND

If your answer is "Yes" to any of the following, explain fully in a signed and notarized statement, including all related details; include the violation, location, date and disposition. Letters from attorneys or insurance companies are not accepted in lieu of your statement. Falsification of any of the following is grounds for permanent revocation of the permit issued pursuant to this application.

| | | |
|---|---|---|
| 1. Has disciplinary action ever been taken regarding any license, certificate, registration or permit you hold or have held? | ☐ Yes | ☐ No |
| 2. Have you ever been denied a license, certificate, registration or permit in any state (including Indiana)? | ☐ Yes | ☐ No |
| 3. Except for minor violations of traffic laws resulting in fines, and arrests or convictions that have been expunged by a court,  (1) have you ever been arrested; | ☐ Yes | ☐ No |
| (2) have you ever entered into a prosecutorial diversion or deferment agreement regarding any offense, misdemeanor, or felony in any state; | ☐ Yes | ☐ No |
| (3) have you ever been convicted of any offense, misdemeanor, or felony in any state; | ☐ Yes | ☐ No |
| (4) have you ever pled guilty to any offense, misdemeanor, or felony in any state; or | ☐ Yes | ☐ No |
| (5) have you ever pled nolo contendre to any offense, misdemeanor, or felony in any state? | ☐ Yes | ☐ No |
| 4. Do you have any condition or impairment (including a history of alcohol or substance abuse) that currently interferes, or if left untreated may interfere, with your ability to practice in a competent and professional manner? | ☐ Yes | ☐ No |

### APPLICATION AFFIRMATION

I hereby swear or affirm, under the penalties of perjury, that the statements made in this application are true, complete, and correct.

| Signature of applicant | Date signed (month, day, year) |
|---|---|

**Figure 25-1**   Indiana application form (*continued*)

| SECTION 2 - INDIANA RESIDENTS |
|---|

*INSTRUCTIONS:*
*For applicants that have completed a four (4) year apprenticeship in an approved program.*

*To be completed by the approved Apprentice School Official.*
- *Applicants who qualify under this section must complete sections 1 and 2 of this application. No other sections are required.*

I have successfully completed the following four (4) years of training and successfully passed a practical examination in an approved apprenticeship program satisfying the requirements as defined in commission rule 860 IAC 1-1-9 and 860 IAC 2-1-7.1.

| Name of apprenticeship program sponsor | Telephone number ( ) |
|---|---|

Address (*number and street, city, state, ZIP code, and county*)

| Date of enrollment (*month, year*) | Date of completion (*month, year*) |
|---|---|

| APPROVED APPRENTICESHIP PROGRAM SPONSOR CERTIFICATION OF COMPLETION |
|---|

I hereby certify that _____
                                                   *Name of apprentice*

successfully completed (4) years of training and successfully passed a practical examination in an approved apprenticeship program, per 860 IAC 1-1-9

and 860 IAC 2-1-7.1.

| Date of enrollment (*month, year*) | Signature of manager of approved apprenticeship program sponsor |
|---|---|
| Date of completion (*month, year*) | Date signed (*month, day, year*) |

| SECTION 3 - FOR OUT-OF-STATE |
|---|

*INSTRUCTIONS:*
*For licensed Journeyman Plumbers applying for Indiana Journeyman Plumber license or non-licensed applicants applying for Indiana Journeyman Plumber license.*

*To be completed by an out-of-state Journeyman Plumber:*
- *Applicant must submit a certified copy of their current license(s) issued by another state.*
- *Applicants are required to have a minimum of four (4) years of license plumbing experience as defined in IC 25-28.5-1-12 and 860 IAC 1-1-11.*
- *Employer's license number is required on application per 860 IAC 1-1-10.*
- *Applicants who qualify under this section must complete section 1 and 3 of this application. No other sections are required.*

*To be completed by non-licensed and nonregistered individuals who are applying by the trade experience, exceptions allowed in IC 25-28.5-1-32 subsections (2), (6), or (7).*
- *Applicant must submit a notarized affidavit prepared by their employer verifying the applicants plumbing trade experience, duties, and employment dates. Applicants who are unable to obtain employer affidavit must submit a notarized affidavit stating the inability and reason why.*
- *The Indiana Plumbing Commission under the authority of 860 IAC 1-1-10 may, after review of the application, request additional information or supporting documents.*
- *Applicants are required to have a minimum of four (4) years of exempt plumbing trade experience.*
- *Applicants who qualify under this section must complete sections 1 and 3 of this application. No other sections are required.*

I have completed the following four (4) years of experience in the plumbing trade, satisfying the requirements as defined in commission rule, 860 IAC 1-1-9 and 860 IAC 1-1-10, as verified by employer, attached herewith:

| Name of employer | Plumbing contractor license number (*if applicable*): PC |
|---|---|
| Address (*number and street, city, state, and ZIP code*) | |

| County | Telephone number ( ) | Dates of employment (*month, day, year*) From              To |
|---|---|---|

| Name of employer | Plumbing contractor license number (*if applicable*): PC |
|---|---|
| Address (*number and street, city, state, and ZIP code*) | |

| County | Telephone number ( ) | Dates of employment (*month, day, year*) From              To |
|---|---|---|

**Figure 25-1**   Indiana application form (*continued*)

| EMPLOYER AFFIDAVIT OF EXPERIENCE IN PLUMBING TRADE | | |
|---|---|---|

I hereby certify that _____ has worked in the plumbing business as
*Name of applicant*

defined in commission rule 860 IAC 1-1-10 for the period of _____ to _____.
*Date (month, day, year)*      *Date (month, day, year)*

| Signature of employer or licensed plumbing contractor | Name of company or plumbing business | Plumbing contractor license number |
|---|---|---|
| Address *(number and street, city, state, and ZIP code)* | | Date signed *(month, day, year)* |

**Licensees who submit false information may be subject to disciplinary action by the Indiana Plumbing Commission.**

| NOTARY CERTIFICATE | | |
|---|---|---|

STATE OF _____

COUNTY OF _____  } SS:

I, _____, having been duly sworn on oath, say that I am the above-named,
that I have personally prepared the foregoing affidavit, and that the same is true to the best of my knowledge and belief.

| Signature of employer | Signature of Notary Public | |
|---|---|---|
| Printed or typed name of employer | Printed or typed name of Notary Public | |
| Date subscribed and sworn to Notary Public *(month, day, year)* | County of residence | Date commission expires *(month, day, year)* |

**PLEASE TAPE YOUR PHOTOGRAPH BELOW *(DO NOT STAPLE)*.**
*(You must place your signature on the front of your photograph.)*

**Figure 25-2**   Michigan application form

**Registration for Plumbing Apprentice**    **83**
Michigan Department of Licensing and Regulatory Affairs
Bureau of Construction Codes / Plumbing Division
P.O. Box 30255, Lansing, MI 48909
517-241-9330
www.michigan.gov/bcc

| Agency Use Only |
|---|
| 83- _____ |
| Batch ____ ____ ____ 114 |
| Date _____ |

Fee: $15.00

| | |
|---|---|
| Authority:  2002 PA 733<br>Completion:  Mandatory<br>Penalty:  Certificate of Registration will not be issued | LARA is an equal opportunity employer/program. Auxiliary aids, services and other reasonable accommodations are available upon request to individuals with disabilities. |

**Instructions** - This form must be submitted within 30 days of employment as a plumbing apprentice. The master plumber having supervision **shall** sign the application and provide his/her license number.

• Complete and sign application. Type or print in ink.
• 1996 PA 236, as amended requires an applicant to include his or her social security number. However, a requirement under this section to include a social security number on an application does not apply to an applicant who demonstrates he or she is exempt under law from obtaining a social security number or to an applicant who for religious convictions is exempt under law from disclosure of his or her social security number under these circumstances.
• Enclose a check made payable to the **State of Michigan**.
• Mail completed application (**all pages must be submitted**) and payment to the address listed above.

**Applicant Information**

| NAME (Last, First, Middle) | DATE OF BIRTH | AGE |
|---|---|---|
| HOME ADDRESS | SOCIAL SECURITY NUMBER* | DATE APPRENTICESHIP BEGAN |

| CITY | STATE | ZIP CODE | COUNTY | TELEPHONE NUMBER (Include Area Code) |
|---|---|---|---|---|

**Apprenticeship School**

Have you attended an apprenticeship school?    ☐ Yes (Complete information below)    ☐ No

| NAME OF SCHOOL | INSTRUCTOR | DATES (MO/DAY/YR) |
|---|---|---|
| | | FROM:          TO: |

**Education**

| HIGH SCHOOL | HIGHEST GRADE COMPLETED | DATE GRADUATED |
|---|---|---|
| COLLEGE/UNIVERSITY | MAJOR | DATE GRADUATED |

**Employment**

| PRESENT EMPLOYER | NAME OF MASTER PLUMBER |
|---|---|

| BUSINESS ADDRESS (Street No. and Name) | CITY | STATE | ZIP CODE |
|---|---|---|---|

DATES OF EMPLOYMENT (MO/DAY/YR)
From:                    To:

**Signature of Master Plumber Having Supervision Responsibility**

I certify the applicant is employed by the above named company for which I am the authorized master plumber. I further understand falsification of any statement is cause for rejection of application or revocation of license, if issued.

| SIGNATURE OF MASTER PLUMBER |
|---|

| LICENSE NUMBER | DATE |
|---|---|

*This information is confidential. Disclosure of confidential information is protected by the Federal Privacy Act.

**Figure 25-2** Michigan application form (*continued*)

**Background Information**

| Have you been convicted of a felony or misdemeanor?  ☐ Yes   ☐ No |
| If yes, complete the Conviction History section below. Failure to accurately respond to this question will result in you forfeiting any rights of consideration for examination and issuance of a plumber's license in the state of Michigan. |

**Conviction History**

In accordance with the Former Offenders Act, 1974 PA 381, this is to provide you with an opportunity to explain your affirmative response to the question above which asked if you had been convicted of a felony or misdemeanor.

If you are unsure of exact details, respond to the best of your knowledge. The information requested on this form is required under 2002 PA 733 and will be used to process your application. Attach additional sheet(s) if necessary.

| YOUR NAME WHEN CONVICTED |
| --- |
| |
| INDICATE CONVICTION(S) FOR WHICH YOU WERE CHARGED |
| |
| DATE(S) OF CONVICTION(S) AND SENTENCE(S) |
| |
| NAME AND ADDRESS OF SENTENCING COURT(S) |
| |
| CHECK YES OR NO TO THE FOLLOWING |
| 1.  Are you a current inmate?                    ☐ Yes   ☐ No |
| 2.  Are you currently on probation / parole?     ☐ Yes   ☐ No |
| If yes, provide the name, address and telephone number of the correctional facility, probation officer or parole officer. |
| |
| RELEASE DATE FROM CUSTODY, PROBATION, OR PAROLE |
| |
| REHABILITATION PROGRAMS ENROLLED IN OR COMPLETED |
| |

**Conviction History Certification and Signature** (To be signed only if Conviction History section above is completed)

| I hereby certify the statements and facts provided are true and accurate to the best of my knowledge. By signing this form, I give my permission to allow the Bureau of Construction Codes to contact appropriate agencies regarding my record of conviction(s). | |
| --- | --- |
| SIGNATURE OF APPLICANT | DATE |

**Certification and Signature (Must** be signed by all applicants)

| I certify all information provided is true and accurate to the best of my ability. I further understand falsification of any statement is cause for rejection of application or revocation of plumbing apprentice registration, if issued. | |
| --- | --- |
| SIGNATURE OF APPLICANT | DATE |

BCC-3328 (Rev. 12/12) Back

## Nevada (and Las Vegas)

The state of Nevada requires the passing of two exams, as well as a business and law exam. Las Vegas, Nevada, in Clark County, is an example of a larger city requiring licenses for both electricians and plumbers.

## New Jersey

In New Jersey, the master plumber must complete five credits of continuing education every biennial renewal period. The State Board of Examiners of Master Plumbers is required to determine the topics required for each biennial renewal period and to put these topics in the New Jersey Register. In the 2003 to 2005 biennial renew period, the Board decided to require every licensed master plumber to complete one hour of review of the existing National Standard Plumbing Code and one hour reviewing the laws and regulations pertinent to the practice of a master plumber (specifically, NJSA 45:14C-l et seq. and NJAC 13:32). The remaining three hours will be devoted to one or more of the following:

- Further review of existing National Standards Plumbing Code
- Further review of laws and regulations pertinent to the practice of a master plumber
- Job site safety, including the Occupational Safety and Health Administration (OSHA) and related information
- American Disabilities Act (ADA) barrier-free provisions and Uniform Construction Code (UCC) requirements
- Plumbing systems pipe sizing for waste, drainage, vent, water, and gas
- Contract liability, insurance requirements, business law, consumer contracts, and ethical conduct
- The licensed master plumber's responsibilities for an employee's workmanship and competence

- International Mechanical Code requirements
- Back-flow prevention and back-flow devices
- New Jersey sales tax and other tax matters
- Organizing/managing a business, estimating/bidding, and project managing
- Excavation and backfill requirements and procedures

## New Mexico

In New Mexico, the Contractor's Licensing Service Inc., provides a free newsletter, services, books, and programs for obtaining the plumber's license. These programs (as well as official programs in technical schools and two-year colleges) are available from the Plumbing License Bureau or Division in each state. Licensing comes under the Attorney General's office in some states, the State Plumbing Board or State's Plumbing Bureau or the State Plumbing Commission. The Internet is a good place to look for the proper licensing authority in any state. Contractor, master, journeyman, and apprentice applications are available from these licensure boards. They also have a schedule of when and where the examinations are given and the fee for each. It is best to know what class of license is being sought at the time of application.

New Mexico has many journeyman and contractor license classifications. New Mexico requires the passing of two exams, as well as a business and law exam.

## South Dakota

The state of South Dakota's State Plumbing Commission has as its mission statement the following:

To protect the public from unsafe drinking water and unsafe waste disposal facilities by licensing qualified plumbers; to inspect plumbing installations and ensure that the state plumbing code is updated and distributed; to inform plumbers, inspection departments and the public of code requirements, new products and methods of installation; and to utilize seminars and the media to provide information of the board's activities, recommendations and requirements.

The South Dakota State Plumbing Commission is responsible for granting plumbing licenses to qualified candidates. The State Plumbing Commission administers and establishes educational, training, and examination requirements for the plumbing contractor, journeyman, and apprentice; sewer and water contractor, installer, and apprentice; appliance contractor, installer, and apprentice; water conditioning contractor, installer, and apprentice; mobile home contractor, installer, and apprentice; and underground irrigation contractor, installer, and apprentice. Plumbing apprentices must serve a four-year apprenticeship. All other apprentices must serve a two-year apprenticeship. Journeyman plumbers must work for two years before becoming a plumbing contractor. All installers must work for one year to become a contractor.

Exams are held in Pierre at any time by advance appointment. Exams are also held in Sioux Falls and Rapid City but not on specific, preset dates. Exams are held in these locations when the office has received enough applications to warrant travel to those sites. Eight to ten applicants are minimum.

## Texas

In Texas, the license is for journeyman, master, and, if an electrician, it can also be for signs. Air-conditioning licenses in Texas may be divided into A and B. The A license is unlimited tonnage of air conditioning, while the B license is limited to work with 15 tons or less.

## Utah

In Utah, the statewide plumbing license is for residential, journeyman, and master.

Utah requires the passing of **two** exams, as well a **business** and **law** exam.

# Application Form

The South Dakota application form for the plumber, contractor, installer, and apprentice is shown in Figure 25-3 for easy reference and guidance purposes. It should give you an idea as to what is needed to apply for a license or apprenticeship.

# Fees

A schedule of fees for first-time exams is listed in Table 25-3. All retakes of the exam are $50.

**Table 25-3**   Schedule of Fees in South Dakota Exam

| | |
|---|---|
| Appliance installation contractor | $175 |
| Appliance installer | $100 |
| Irrigation contractor | $175 |
| Irrigation installer | $100 |
| Journeyman plumber | $130 |
| Mobile home contractor | $175 |
| Mobile home installer | $100 |
| Plumbing contractor | $260 |
| Sewer and water contractor | $225 |
| Sewer and water installer | $100 |
| Temporary permit | $50 |
| Water conditioning contractor | $175 |
| Water conditioning installer | $100 |

**Figure 25-3**   South Dakota application form

SD EForm - 0258   V5

NOTE: FAILURE TO COMPLY WITH ALL INSTRUCTIONS WILL CAUSE APPLICATION TO BE RETURNED!

# SOUTH DAKOTA STATE PLUMBING COMMISSION
308 South Pierre Street
c/o 1320 East Sioux Avenue
Pierre, South Dakota 57501-2017
Phone: (605) 773-3429
FAX: (605) 773-5405

## INSTRUCTIONS FOR COMPLETING THIS APPLICATION

1. This application must be typewritten or printed in ink.
2. Complete all spaces provided. If the Question does not apply, write "none" in the space.
3. Sign the application.
4. Accompany this application with the appropriate fee as indicated on the application.
5. Accompany this application with written statements from present and previous employers which must state: a) dates of employment; b) number of hours worked during employment; and, c) extent of work performed during employment.
6. Reciprocity applicants – accompany this application with a photo-copy of your current license from the state in which you are licensed and disregard instruction #5. (MN, ND, MT, CO RESIDENTS ONLY)
7. Applicants for apprentice licenses may disregard instructions #5 and #6.

NOTE: FAILURE TO COMPLY WITH ALL INSTRUCTIONS WILL CAUSE APPLICATION TO BE RETURNED!

| | | | |
|---|---|---|---|
| ☐ PLUMBING CONTRACTOR | $ 340.00 | ☐ WATER CONDITIONING CONTRACTOR | $ 240.00 |
| ☐ JOURNEYMAN PLUMBER | $ 185.00 | ☐ WATER CONDITIONING INSTALLER | $ 155.00 |
| ☐ APPRENTICE PLUMBER | $ 10.00 | ☐ W/C INST. APPRENTICE | $ 10.00 |
| (Less than 4 years) | | (Less than 2 years) | |
| ☐ APPLIANCE INSTALLATION CONTRACTOR | $ 240.00 | ☐ SEWER & WATER CONTRACTOR | $ 300.00 |
| ☐ APPLIANCE INSTALLATION INSTALLER | $ 155.00 | ☐ SEWER & WATER INSTALLER | $ 155.00 |
| ☐ APPL. INST. APPRENTICE | $ 10.00 | ☐ S. & W. INST. APPRENTICE | $ 10.00 |
| (Less than 2 years) | | (Less than 2 years) | |
| ☐ MOBILE HOME CONTRACTOR | $ 240.00 | ☐ IRRIGATION CONTRACTOR | $ 240.00 |
| ☐ MOBILE HOME APPRENTICE | $ 10.00 | ☐ IRRIGATION INSTALLER | $ 155.00 |
| ☐ MOBILE HOME INSTALLER | $ 155.00 | ☐ IRRIGATION APPRENTICE | $ 10.00 |
| ☐ TEMPORARY PERMIT | $ 50.00 | ☐ CODE BOOK | $ 140.00 |
| | | ☐ STUDY GUIDE | $ 70.00 |
| | | ☐ UTILITY HANDBOOK | $ 50.00 |

NAME _____  DATE _____ (MM/DD/YYYY)

SOCIAL SECURITY NUMBER _____  RES. PHONE NUMBER _____

MAILING ADDRESS _____
Street Number   County   City   State   Zip Code

YOUR AGE ____  YOUR DATE OF BIRTH _____ (MM/DD/YYYY)  YOUR PLACE OF BIRTH _____ City   State

PRESENT EMPLOYER _____  WORK PHONE NUMBER _____

EMPLOYED AS _____

ADDRESS OF EMPLOYER _____

PB201B Rev: 12-02-500

| | |
|---|---|
| | NEED BOARD SIGNING |
| | SEND REFERENCES ( ) |

**Figure 25-3**   South Dakota application form (*continued*)

Have you ever carried a Plumbing License?  Y◯ N◯        If so, where?_____

State the type or grade of License _____ In force, from _____ to _____

Was the License obtained by examination? Y◯ N◯ Have you ever had a Plumbing License revoked? Y◯ N◯ By whom?_____
If so, give reasons _____

Have you previously been examined for a Plumbing License by this commission?   Y◯ N◯
If so, state type, and results of examination _____  Approved ☐  Disapproved ☐

Have you previously made application for a State of South Dakota Plumbing License?  Y◯ N◯

Is your spouse an active duty member of the armed forces?  Y ◯ N ◯

If Yes, is your spouse subject to military transfer to South Dakota, and did you leave employment to accompany your spouse to South Dakota?  Y ◯ N ◯

## SCHOOL RECORD

Education:

Highest Grade Completed   1☐ 2☐ 3☐ 4☐ 5☐ 6☐ 7☐ 8☐ 9☐ 10☐ 11☐ 12☐ 13☐ 14☐ 15☐ 16☐

Are you a graduate of a Plumbing Course of an accredited University or College?  Y◯ N◯

Give degree _____ Year _____ Name of School _____

Address of School _____

Are you a graduate of a Plumbing Trade School?  Y◯ N◯

Name of above School _____

Address of above School _____

State other courses of Plumbing Study, if any _____

Name and address of above _____

## EMPLOYMENT DATA

Be sure that you break down your experience according to each classification.

| Total years of Plumbing Experience | EXPERIENCE | | | | | |
|---|---|---|---|---|---|---|
| | As Apprentice | | As Journeyman | | As Contractor | |
| CLASSIFICATION | Months | Years | Months | Years | Months | Years |
| Residential Plumbing | 0.0 | 0.0 | 0.0 | 0.0 | 0.0 | 0.0 |
| Commercial & Industrial Plumbing | 0.0 | 0.0 | 0.0 | 0.0 | 0.0 | 0.0 |
| Farmstead Plumbing | 0.0 | 0.0 | 0.0 | 0.0 | 0.0 | 0.0 |
| Plumbing Maintenance & Repair | 0.0 | 0.0 | 0.0 | 0.0 | 0.0 | 0.0 |
| Sewer & Water Installation | 0.0 | 0.0 | 0.0 | 0.0 | 0.0 | 0.0 |
| Appliance Installation | 0.0 | 0.0 | 0.0 | 0.0 | 0.0 | 0.0 |
| Water Cond't. Installation | 0.0 | 0.0 | 0.0 | 0.0 | 0.0 | 0.0 |
| Planning & laying out for | 0.0 | 0.0 | 0.0 | 0.0 | 0.0 | 0.0 |
| Mobile Home Plumbing Work | 0.0 | 0.0 | 0.0 | 0.0 | 0.0 | 0.0 |
| | TOTAL YEARS | 0.00 | TOTAL YEARS | 0.00 | TOTAL YEARS | 0.00 |

**Figure 25-3** South Dakota application form (*continued*)

## REFERENCES

List at least two (2) persons actively engaged in the plumbing industry that you have worked under.

| Name _____ | Name _____ |
|---|---|
| Address _____ | Address _____ |
| Occupation _____ | Occupation _____ |
| Name _____ | Name _____ |
| Address _____ | Address _____ |
| Occupation _____ | Occupation _____ |

## PLUMBING EMPLOYMENT RECORD

| IMPORTANT<br>Unless complete address of employer is given, it is impossible to properly process your application and will cause delay.<br>PREVIOUS AND PRESENT EMPLOYERS | DATES EMPLOYED | | TYPE OF PLUMBING WORK |
|---|---|---|---|
| | From<br>Month  Year | To<br>Month  Year | |
| Name<br>Address | | Present | |
| Name<br>Address | From | To | |
| Name<br>Address | From | To | |

I declare and affirm under the penalties of perjury that this claim (petition, application, information) has been examined by me, and to the best of my knowledge and belief, is in all things true and correct.

_____
SIGNATURE

The disclosure of the applicant's Social Security number on the front page of this application form is mandatory pursuant to 42 USCA 666, Title IV-D of the Social Security Act. The Plumbing Commission will keep the applicant's Social Security number confidential, except that the number may be provided to the Department of Social Services for use in administering Title IV-D of the Social Security Act.

**Figure 25-3**　South Dakota application form (*continued*)

## REMARKS BY APPLICANT

## SPACE BELOW RESERVED FOR COMMISSION

Approved_____ Disapproved_____ for examination.  Types of examination_____ Date_____

By _____ and _____

　　　　　　　Contractor Member　　　　　　　　　　　　　　　　　Plumber Member

| RECORD | Verified | Not Verified | Corrected by _____ | #1 |
|---|---|---|---|---|
| | | | | #2 |
| Training | | | Corrected by _____ | #3 |
| Work Experience | | | | #4 |
| Work Reference | | | License #_____ as _____ | #5 |

Examination   #1   2   3   4   5   Date of Examination_____ Exam Supervised By_____

| RATE AS _____ A<br><br>By_____ Date_____ | DATE_____<br>GRADE_____<br>PASSED　　　FAILED<br>BY_____ | Notes:_____<br>_____<br>_____<br>_____<br>_____ |
|---|---|---|

LICENSE NUMBER ISSUED_____
DATE_____ INITIALS_____

# State Requirements

Keep in mind the fees in any state may change. The ones here are presented to give you an idea of the various fees at this time. To keep up with the existing fees you can go on the Internet and obtain the correct amounts. Requirements for becoming each of the listed installers, contractors, and plumbers are given in the South Dakota State Plumbing Commission's website (www.state.sd.us/dol/boards/plumbing/licens.htm). For those without access to the Internet, the following sections are presented with the Commission's permission.

# Becoming a Plumbing Contractor/Journeyman Plumber/Apprentice

Plumbing contractors, journeyman plumbers, and apprentice plumbers may engage in the furnishing and/or the use of materials and fixtures in the installation, extension, and alteration of all piping, fixtures, appliances, and appurtenances in connection with sanitary drainage or storm drainage facilities, the venting system and the public water supply systems, within or adjacent to any building, structure, or conveyance; and including the installation, extension, or alteration of the storm water, liquid waste, or sewerage and water supply systems of any premises to their connection with any point of public disposal, or other regulated terminal.

Applicants for a plumbing contractor license shall show evidence of six years of experience as a plumbing contractor, plumber, or plumber's apprentice with at least two of those years as a plumbing contractor or plumber.

During the six years, the applicant must have spent at least 1,900 hours per year as a plumbing contractor, plumber, or plumber's apprentice.

Applicants for a plumbing contractor's license must fill out an application and pay an examination and license fee of $260.

Applicants for a journeyman plumber's license shall show evidence of four years of experience as an apprentice plumber.

During the four years, the applicant must have spent at least 1,900 hours per year as an apprentice.

Credit for military plumbing is given at the rate of one year credit for each two years in the military, up to a maximum of five years of credit.

Applicants for a journeyman plumber's license must fill out an application and pay an examination and license fee of $130.

If the commission finds the individuals have the required experience, they shall be tested. Applicants for a plumber apprentice license must fill out an application showing the plumber under which the applicant is working.

The license is without charge for the first four years and then the fee is $80.

Apprentice plumbers who have had three years (5,700 hours) experience in learning and assisting in the installation, alteration, and repair of plumbing under a plumbing contractor may work during their fourth year of apprenticeship without being under the direct supervision of a plumbing contractor or plumber.

# Becoming a Sewer and Water Contractor/ Installer/Apprentice

Sewer and water contractors, installers, and apprentices may engage in the setting up of building sewer and water service, the repair of existing building sewer and water services, the setting up of building storm sewers, the repair of existing building storm sewers, and the setting up of water treatment plant piping and equip-

ment designed to purify water, chemical treatment piping, and sewer treatment plant piping
and equipment designed to treat sewage, and the
repair of the piping and equipment.

- Applicants for a sewer and water contractor
  license shall show evidence of one year of
  experience as a sewer and water installer.
- Applicants for a sewer and water installer
  license shall show evidence of two years of
  experience as a sewer and water apprentice.
- If the commission finds the individuals have
  the required experience, they shall be tested.
- Applicants must fill out an application form.
- A sewer and water contractor shall pay an
  examination and license fee of $225.
- A sewer and water installer shall pay an
  examination and license fee of $100.
- All applicants for a sewer and water apprentice license shall fill out an application
  showing the sewer and water installer under
  which the applicant is working. The license
  fee is without charge for two years, and then
  the fee is $50.

## Becoming an Appliance Contractor/Installer/ Apprentice

Appliance contractors, installers, and apprentices may engage in the making of local connections to water and waste systems of all miscellaneous and commercial equipment designed to
use electricity or gas (or both) as a basic source
of energy; maintenance and service of this equipment, including operation, adjustment, repair,
removal, and renovation; but not connection
and repair of household and commercial plumbing fixtures that are designed for sanitary use or
water conditioning.

- Applicants for an appliance contractor
  license shall show evidence of one year of
  experience as an appliance installer.

- Applicants for an appliance installer license
  shall show evidence of two years of experience as an appliance apprentice.
- If the commission finds the individuals have
  the required experience, they shall be tested.
- Applicants must fill out an application form.
- An appliance contractor shall pay an examination and license fee of $175.
- An appliance installer shall pay an examination and license fee of $100.
- All applicants for an appliance apprentice
  license shall fill out an application showing the appliance installer under which the
  applicant is working. The license is without
  charge for two years and then the fee is $50.

## Becoming a Water Conditioning Contractor/ Installer/Apprentice

Water conditioning contractors, installers, and
apprentices may engage in the treatment of
water and the installation of appliances, appurtenances, fixtures, and plumbing necessary
thereto, all designed to treat water to alter, mod-
ify, add, or remove mineral, chemical, or bacterial content, and to repair such equipment to a
water distribution system. Water conditioning
installation, repair, and treatment does not mean
the exchange of appliances, appurtenances, and
fixtures when the plumbing has previously been
installed or adapted for such appliances, appurtenances, and fixtures, and no substantial change
in the plumbing system is required.

- Applicants for a water conditioning contractor license shall show evidence of one
  year of experience as a water conditioning
  installer.
- Applicants for a water conditioning installer
  license shall show evidence of two years of
  experience as a water conditioning apprentice. If the commission finds the individuals

have the required experience, they shall be tested.

- Applicants must fill out an application form.
- A water conditioning contractor shall pay an examination and license fee of $175.
- A water conditioning installer shall pay an examination and license fee of $100.
- All applicants for a water conditioning apprentice license shall fill out an application showing the water conditioning installer under which the applicant is working. The license fee is without charge for two years and then the fee is $50.

## Becoming a Mobile Home Contractor/Installer/Apprentice

Mobile home contractors, installers, and apprentices may engage in the connection to local water and waste systems from manufactured and mobile homes only, including maintenance and service (which includes the operation, adjustment, repair, removal and renovation of such connections).

- Applicants for a mobile home contractor license shall show evidence of one year of experience as a mobile home installer.
- Applicants for a mobile home installer license shall show evidence of two years of experience as a mobile home apprentice.
- If the commission finds the individuals have the required experience, they shall be tested.
- Applicants must fill out an application form.
- A mobile home contractor shall pay an examination and license fee of $175.
- A mobile home installer shall pay an examination and license fee of $100.
- All applicants for a mobile home apprentice license shall fill out an application showing

the mobile home installer under which the applicant is working. The license is without charge for two years and then the fee is $50.

## Becoming an Underground Irrigation Contractor/Installer/Apprentice

Underground irrigation contractors, installers, and apprentices may engage in the practice of and the furnishing or use of materials and devices for the purpose of installing underground irrigation systems and connecting them to the local source of potable water, including maintenance and service (which includes the operation, adjustment, repair, removal, and renovation of such connections and devices).

- Applicants for an underground irrigation contractor license shall show evidence of one year of experience as an underground irrigation installer.
- Applicants for an underground irrigation installer license shall show evidence of two years of experience as an underground irrigation apprentice.
- If the commission finds the individuals have the required experience, they shall be tested.
- Applicants must fill out an application form.
- An underground irrigation contractor shall pay an examination and license fee of $175.
- An underground irrigation installer shall pay an examination and license fee of $100.
- All applicants for an underground irrigation apprentice license shall fill out an application showing the underground irrigation installer under which the applicant is working. The license is without charge for two years and then the fee is $50.

# Review Questions

1. Why do individual states have licensing requirements for plumbers?

   _____

2. What is an apprentice?

   _____

3. How long are apprenticeships?

   _____

4. How do you become an apprentice?

   _____

5. How do you become a journeyman?

   _____

6. How do you become a contractor?

   _____

7. Does reciprocity exist for journeyman and contractor plumbers?

   _____

8. Which states require the passing of two exams for a license to work as a plumber?

   _____

9. How does the American Disabilities Act concern plumbers?

   _____

10. Where are the use of backflow devices usually encountered?

    _____

# A | Techniques for Studying and Test-Taking

## Preparing for the Exam

1. **Make a study schedule.** Assign yourself a period of time each day to devote to preparation for your exam. A regular time is best, but the important thing is daily study.
2. **Study alone.** You will concentrate better when you work by yourself. Keep a list of questions you find puzzling and points you are unsure of to talk over with a friend who is preparing for the same exam. Plan to exchange ideas at a joint review session just before the test.
3. **Eliminate distractions.** Choose a quiet, well-lit spot as far as possible from the telephone, television, and family activities. Try to arrange not to be interrupted.
4. **Begin at the beginning.** Read. Underline points that you consider significant. Make marginal notes. Flag the pages that you think are especially important with little Post-it notes.
5. **Concentrate on the information and instruction chapters.** Study the Code Definitions, the Dictionary of Plumbing and Terms, and the Dictionary of Equipment and Usage. Learn the language of the field. Focus on the technique of eliminating wrong answers. This information is important to answering all multiple-choice questions.
6. **Answer the practice questions chapter by chapter.** Take note of your weaknesses; use all available textbooks to brush up.
7. **Try the previous exams, if available.** When you believe that you are well prepared, move on to these exams. If possible, answer an entire exam in one sitting. If you must divide your time, divide it into no more than two sessions per exam.

When you do take the practice exams, treat them with respect. Consider each as a dress rehearsal for the real thing. Time yourself accurately, and do not peek at the correct answers.

Remember, you are taking these for practice. They will not be scored; they do not count. So learn from them.

### IMPORTANT

Do not memorize questions and answers. Any question that has been released will not be used again. You may run into questions that are very similar, but you will not be tested with these exact questions. These questions will give you good practice, but they will not have the exact answers to any of the questions on your exam.

## How to Take an Exam

- **Get to the examination room about 10 minutes ahead of time.** You'll get a better start when you are accustomed to the room. If the room is too cold, too warm, or not well ventilated, call these conditions to the attention of the person in charge.
- **Make sure that you read the instructions carefully.** In many cases, test takers lose points because they misread some important part of the directions. (An example would be reading the incorrect choice instead of the correct choice.)
- **Don't be afraid to guess.** The best policy is, of course, to pace yourself so that you can read and consider each question. Sometimes this does not work. Most civil service exam scores are based only on the number of questions answered correctly. This means that a wild guess is better than a blank space. There is no penalty for a wrong answer, and you just might guess right. If you see that time is about to run out, mark all the remaining spaces with the same answer. According to the law of averages, some will be right.

However, you have bought this book for practice in answering questions. Part of your preparation is learning to pace yourself so that you need not answer randomly at the end. Far better than a wild guess is an educated guess. You make this kind of guess not when you are pressed for time, but when you are not sure of the correct answer. Usually, one or two of the choices are obviously wrong. Eliminate the obviously wrong answers and try to reason among those remaining. Then, if necessary, guess from the smaller field. The odds of choosing a right answer increase if you guess from a field of two instead of from a field of four. When you make an educated guess or a wild guess in the course of the exam, you might want to make a note next to the question number in the test booklet. Then, if there is time, you can go back for a second look.

■ Reason your way through multiple-choice questions very carefully and methodically.

# Multiple-Choice Test-Taking Tips

Here are a few examples that we can "walk through" together:

1. On the job, your supervisor gives you a hurried set of directions. As you start your assigned task, you realize you are not quite clear on the directions given to you. The best action to take would be to:
   (A) continue with your work, hoping to remember the directions.
   (B) ask a co-worker in a similar position what he or she would do.
   (C) ask your supervisor to repeat or clarify certain directions.
   (D) go on to another assignment.

   In this question you are given four possible answers to the problem described. Though the four choices are all possible actions, it is up to you to choose the best course of action in this particular situation.

Choice (A) will likely lead to a poor result; given that you do not recall or understand the directions, you would not be able to perform the assigned task properly. Keep choice (A) in the back of your mind until you have examined the other alternatives. It could be the best of the four choices given.

Choice (B) is also a possible course of action, but is it the best? Consider that the co-worker you consult has not heard the directions. How could he or she know? Perhaps his or her degree of incompetence is greater than yours in this area. Of choices (A) and (B), the better of the two is still choice (A).

Choice (C) is an acceptable course of action. Your supervisor will welcome your questions and will not lose respect for you. At this point, you should hold choice (C) as the best answer and eliminate choice (A).

The course of action in choice (D) is decidedly incorrect because the job at hand would not be completed. Going on to something else does not clear up the problem; it simply postpones your having to make a necessary decision.

After careful consideration of all choices given, choice (C) stands out as the best possible course of action. You should select choice (C) as your answer.

Every question is written about a fact or an accepted concept. The question above indicates the concept that, in general, most supervisory personnel appreciate subordinates' questioning directions that may not have been fully understood. This type of clarification precludes subsequent errors on the part of subordinates. On the other hand, many subordinates are reluctant to ask questions for fear that their lack of understanding will detract from their supervisor's evaluation of their abilities.

The supervisor, therefore, has the responsibility of issuing orders and directions in such a way that subordinates will not be discouraged from asking questions. This is the concept on which the sample question was based.

Of course, if you were familiar with this concept, you would have no trouble answering the question. However, if you were not familiar with it, the method outlined here of eliminating incorrect choices and selecting the correct one should prove successful for you.

We have now seen how important it is to identify the concept and the key phrase of the question. Equally, or perhaps even more important, is identifying and analyzing the keyword or the qualifying word in a question. This word is usually an adjective or adverb. Some of the most common key words are:

| | | |
|---|---|---|
| most | least | best |
| highest | lowest | always |
| never | sometimes | tallest |
| most likely | greatest | smallest |
| average | easiest | most |
| nearly | maximum | minimum |
| only | chiefly | mainly |
| but | or | |

Identifying these key words is usually half the battle in understanding and, consequently, answering all types of exam questions.

Now we will use the elimination method on some additional questions:

2. On the first day you report for work after being appointed as a plumber's helper, you are assigned to routine duties that seem to you to be very petty in scope. You should:
   (A) perform your assignment perfunctorily while conserving your energies for more important work in the future.
   (B) explain to your supervisor that you are capable of greater responsibility.
   (C) consider these duties an opportunity to become thoroughly familiar with the workplace.
   (D) try to get someone to take care of your assignment until you have become thoroughly acquainted with your new associates.

Once again we are confronted with four possible answers from which we are to select the best one.

Choice (A) will not lead to getting your assigned work done in the best possible manner in the shortest possible time. This would be your responsibility as a newly appointed plumber's helper, and the likelihood of getting to do more important work in the future following the approach stated in this choice is remote. However, since this is only choice (A), we must hold it aside because it may turn out to be the best of the four choices given.

Choice (B) is better than choice (A) because your superior may not be familiar with your capabilities at this point. We therefore should drop choice (A) and retain choice (B) because, once again, it may be the best of the four choices.

The question clearly states that you are newly appointed. Therefore, would it not be wise to perform whatever duties you are assigned in the best possible manner? In this way, you would not only use the opportunity to become acquainted with procedures, but also to demonstrate your abilities.

Choice (C) contains a course of action that will benefit you and the location in which you are working because it will get needed work done. At this point, we drop choice (B) and retain choice (C) because it is by far the better of the two.

The course of action in choice (D) is not likely to get the assignment completed, and it will not enhance your image to your fellow apprentices.

Choice (C), when compared to choice (D), is far better and therefore should be selected as the best choice.

Now let us take a question that appeared on a police officer examination:

3. An off-duty police officer in civilian clothes riding in the rear of a bus notices two teen-age boys tampering with the rear emergency door. The most appropriate action for the officer to take is to:

   (A) tell the boys to discontinue their tam-pering, pointing out the dangers to life that their actions may create.

   (B) report the boys' actions to the bus operator and let the bus operator take whatever action is deemed best.

   (C) signal the bus operator to stop, show the boys the officer's badge, and then order them off the bus.

   (D) show the boys the officer's badge, order them to stop their actions, and take down their names and addresses.

Before considering the answers to this ques-tion, we must accept that it is a well-known fact that a police officer is always on duty to uphold the law even though he or she may be techni-cally off duty.

In choice (A), the course of action taken by the police officer will probably serve to educate the boys and get them to stop their unlawful activity. Since this is only the first choice, we will hold it aside.

In choice (B), we must realize that the authority of the bus operator in this instance is limited. He can ask the boys to stop tampering with the door, but that is all. The police officer can go beyond that point. Therefore, we drop choice (B) and continue to hold choice (A).

Choice (C) as a course of action will not have a lasting effect. What is to stop the boys from boarding the next bus and continuing their unlawful action? We therefore drop choice (C) and continue to hold choice (A).

Choice (D) may have some beneficial effect, but it would not deter the boys from continuing their actions in the future.

When we compare choice (A) with choice (D), we find that choice (A) is the better one overall, and therefore it is the correct answer.

The next question illustrates a type of ques-tion that has gained popularity in recent exam-inations and that requires a two-step evaluation.

First, the reader must evaluate the condition in the question as being "desirable" or "undesir-able." Once the determination has been made, we are then left with making a selection from two choices instead of the usual four.

4. A visitor to an office in a city agency tells one of the office aides that he has an appointment with the supervisor of the office who is expected shortly. The visitor asks for permission to wait in the supervi-sor's private office, which is unoccupied at the moment. For the office aide to allow the visitor to do so would be:

   (A) desirable; the visitor would be less likely to disturb the other employees or to be disturbed by them.

   (B) undesirable; it is not courteous to per-mit a visitor to be left alone in an office.

   (C) desirable; the supervisor may wish to speak to the visitor in private.

   (D) undesirable; the supervisor may have left confidential papers on the desk.

First of all, we must evaluate the course of action on the part of the office aide of permit-ting the visitor to wait in the supervisor's office as being very undesirable. There is nothing said of the nature of the visit; it may be for a pur-pose that is not friendly or congenial. There may be papers on the supervisor's desk that he or she does not want the visitor to see or to have knowledge of. Therefore, at this point, we have to decide between choices (B) and (D).

This is definitely not a question of courtesy. Although all visitors should be treated with courtesy, permitting the visitor to wait in the supervisor's office in itself is not the only possible act of courtesy. Another comfortable place could be found for the visitor to wait.

Choice (D) contains the exact reason for evaluating this course of action as being undesirable, and when we compare it with choice (B), choice (D) is far better.

## A Strategy for Test Day

On the exam day assigned to you, allow the test itself to be the main attraction of the day. Do not squeeze it in between other activities. Arrive rested, relaxed, and on time. In fact, plan to arrive a little bit early. Leave plenty of time for traffic tie-ups or other complications that might upset you and interfere with your test performance.

Here is a breakdown of what occurs on examination day and tips on starting off on the right foot and preparing to start your exam:

1. In the test room, the examiner will hand out forms for you to fill out and will give you the instructions that you must follow in taking the examination. Note that you must follow instructions exactly.
2. The examiner will tell you how to fill in the blanks on the forms.
3. Exam time limits and timing signals will be explained.
4. Be sure to ask questions if you do not understand any of the examiner's instructions. You need to be sure that you know exactly what to do.
5. Fill in the grids on the forms carefully and accurately. Filling in the wrong blank may lead to loss of veterans' credits to which you may be entitled or to an incorrect address for your test results.

6. Do not begin the exam until you are told to begin.
7. Stop as soon as the examiner tells you to stop.
8. Do not turn pages until you are told to do so.
9. Do not go back to parts you have already completed.
10. Any infraction of the rules is considered cheating. If you cheat, your test paper will not be scored, and you will not be eligible for appointment.
11. Once the signal has been given and you begin the exam, read every word of every question.
12. Be alert for exclusionary words that might affect your answer—words like "not," "most" and "least."

## Marking Your Answers

Read all the choices before you mark your answer. It is statistically true that most errors are made when the last choice is the correct answer. Too many people mark the first answer that seems correct without reading through all the choices to find out which answer is best.

Be sure to read the suggestions below now and review them before you take the actual exam. Once you are familiar with the suggestions, you will feel more comfortable with the exam itself and find them all useful when you are marking your answer choices.

1. Mark your answers by completely blackening the answer space of your choice.
2. Mark only *one* answer for each question, even if you think that more than one answer is correct. You must choose only one. *If* you mark more than one answer, the scoring machine will consider you wrong even if one of your answers is correct.
3. *If* you change your mind, erase completely. Leave no doubt as to which answer you have chosen.

4.  If you do any figuring on the test booklet or on scratch paper, be sure to mark your answer on the answer sheet.

5.  Check often to be sure that the question number matches the answer space number and that you have not skipped a space by mistake. If you do skip a space, you must erase all the answers after the skip and answer all the questions again in the right places.

6.  Answer every question in order, but do not spend too much time on any one question. If a question seems to be "impossible," do not take it as a personal challenge. Guess and move on. Remember that your task is to answer correctly as many questions as possible. You must apportion your time so as to give yourself a fair chance to read and answer all the questions. If you guess at an answer, mark the question in the test booklet so that you can find it easily if time allows.

7.  Guess intelligently if you can. If you do not know the answer to a question, eliminate the answers that you know are wrong and guess from among the remaining choices. If you have no idea whatsoever of the answer to a question, guess anyway. Choose an answer other than the first. The first choice is generally the correct answer less often than the other choices. If your answer is a guess, either an educated guess or a wild one, mark the question in the question booklet so that you can give it a second try if time permits.

8.  If you happen to finish before time is up, check to be sure that each question is answered in the right space and that there is only one answer for each question. Return to the difficult questions that you marked in the booklet and try them again. There is no bonus for finishing early so use all your time to perfect your exam paper.

With the combination of techniques for studying and test taking as well as the self-instructional course and sample examinations in this book, you have been given the tools you need to score highly on your exam.

# B | Practice Questions on Plumbing

# Practice Problems: True/False

1. PVC (polyvinyl chloride) PVC-type 1 is strong, rigid and economical.
   ❏ True          ❏ False

2. Use of ABS (acrylonine butadiene styrene) has almost doubled when compared with the PVC in DWV piping system.
   ❏ True          ❏ False

3. CPVC meets national standards for piping of 212°F and water pressures up to 200 psi.
   ❏ True          ❏ False

4. Polypropylene is a heavier material suitable for lower pressure applications up to 212°F.
   ❏ True          ❏ False

5. PVDF is a strong, tough, and abrasive resistant fluorocarbon material.
   ❏ True          ❏ False

6. CPVC polypropylene and PVDF at the same temperature will expand approximately 3¼ inches.
   ❏ True          ❏ False

7. Almost all mobile homes and travel trailers have copper piping.
   ❏ True          ❏ False

8. Two types of plastic pipe and fittings are commonly used in drainage systems: PVC and ABS.
   ❏ True          ❏ False

9. Priming is essential with PVC and CPVC.
   ❏ True          ❏ False

10. A fine-toothed power saw can be used to cut PVC and ABS pipe.
    ❏ True          ❏ False

11. Fifty hours are required for ABS piping systems to stand vented to the atmosphere before pressure testing.
    ❏ True          ❏ False

12. Some pipe lubricants contain compounds that may soften the surface of ABS and PVC pipe.
    ❏ True          ❏ False

13. The industry does not recommend threading ABS and PVC Schedule 40 plastic pipe.
    ❏ True          ❏ False

14. If a lubricant is thought to be needed, you can use ordinary Vaseline or pipe tape.
    ❏ True          ❏ False

15. Soft copper is flexible and comes in 10-foot lengths.
    ❏ True          ❏ False

16. There are three types of fittings for soft copper tubing.
    ❏ True          ❏ False

17. Plastic pipe lighter than Schedule 80 should not be threaded.
    ❏ True          ❏ False

18. Septic tanks are designed to operate full.
    ❏ True          ❏ False

19. Drain fields will become spent in time.
    ❏ True          ❏ False

20. The distance from the inlet invert (or bottom part of the inlet) to the septic tank from the top of the tank is generally 12 inches (305 mm).
    ❏ True          ❏ False

21. Once the septic tank is set and leveled, the tank's drain line is set.
    ❏ True          ❏ False

22. Drain tile is spaced approximately one inch apart.
    ❏ True          ❏ False

23. Septic tanks are usually made of concrete or fiberglass.
❑ True        ❑ False

24. Wells are usually dug by plumbers.
❑ True        ❑ False

25. Cisterns are installed by plumbers.
❑ True        ❑ False

26. A cistern is no longer used for drinking water.
❑ True        ❑ False

27. Each county, state and local jurisdiction has its own rules and regulations for permits to discharge into the city sewers.
❑ True        ❑ False

28. Most public sewers consist of a 6-inch plastic pipe that is lateraled off the main sewer.
❑ True        ❑ False

29. Bituminous fiber pipe is easy to cut and is used everywhere.
❑ True        ❑ False

30. Cast iron pipe is coated with coal tar to keep it from rusting.
❑ True        ❑ False

31. A buildings' sewers may also be made of concrete pipe and fittings.
❑ True        ❑ False

32. Vitrified clay pipe has a glass-like coating.
❑ True        ❑ False

33. VCP pipe is attacked by all sewage acids.
❑ True        ❑ False

34. Building sewers may be made of copper pipe or brass pipe.
❑ True        ❑ False

35. National Primary Drinking Water standards are legally enforceable standards that apply to public water systems.
❑ True        ❑ False

36. Primary systems protect public health by limiting the levels of contaminants in the drinking water.
❑ True        ❑ False

37. All sources of drinking water contain some naturally occurring contaminants.
❑ True        ❑ False

38. The Safe Drinking Water Act requires the EPA to establish and enforce standards for public drinking water.
❑ True        ❑ False

39. About 50 percent of people in the U.S. rely on drinking water from private wells.
❑ True        ❑ False

40. Water mains of the MUD are usually located in front of the house.
❑ True        ❑ False

41. All MUD service requires the plumbing to include a stop valve.
❑ True        ❑ False

42. The Environmental Protection Agency is a state agency for water quality control.
❑ True        ❑ False

43. Working drawings for a building are very simple.
❑ True        ❑ False

44. The American Standard Graphical Symbols for Piping are used in the construction industry so the plumbers can know where the plumbing fixtures go.
❑ True        ❑ False

45. Drafting or mechanical drawing is a form of shorthand.
❑ True        ❑ False

46. One of the first use of symbols was in mathematics.
❑ True        ❑ False

47. To abbreviate means to shorten or make smaller.
    ❏ True          ❏ False

48. The earth's crust is made up of 50 percent aluminum.
    ❏ True          ❏ False

49. The earth's crust is made up of 40 percent calcium.
    ❏ True          ❏ False

50. If an element is metallic, its atoms will show up as a positive charge.
    ❏ True          ❏ False

51. Magnesium in rock form is called dolomite.
    ❏ True          ❏ False

52  Iron has a Latin name of ferrous.
    ❏ True          ❏ False

53. Boron and brass are non-metals.
    ❏ True          ❏ False

54. Magnesium is extracted from sea water.
    ❏ True          ❏ False

55. Non-metals have one or more of the physical properties of metal.
    ❏ True          ❏ False

56. All metals that do not contain iron are referred to as non-ferrous.
    ❏ True          ❏ False

57. The U.S. produces about 90 percent of the aluminum produced in the world.
    ❏ True          ❏ False

58. Aluminum melts at a higher temperature than zinc, tin and lead.
    ❏ True          ❏ False

59. Aluminum can be produced in more forms than any other metal.
    ❏ True          ❏ False

60. Copper is used to make brass and bronze.
    ❏ True          ❏ False

61. Copper compounds are used in paint to protect materials against corrosion.
    ❏ True          ❏ False

62. Copper's name comes from Latin for Cyprian.
    ❏ True          ❏ False

63. The green film that forms on copper is called patina.
    ❏ True          ❏ False

64. The amount of carbon in steel determines its classification.
    ❏ True          ❏ False

65. The symbol for lead is Pb, taken from the Latin that calls lead plumbum.
    ❏ True          ❏ False

66. Pewter is made of lead and tin.
    ❏ True          ❏ False

67. Eating from pewter plates can cause mental illness.
    ❏ True          ❏ False

68. Lead is found in veins and masses in limestone and dolomite.
    ❏ True          ❏ False

69. Magnesium ranks second as the most abundant metal in the earth's crust.
    ❏ True          ❏ False

70. Seawater contains about 1.07 kilograms of magnesium per cubic meter (1 pound per 15 cu ft).
    ❏ True          ❏ False

71. The most useful form of mercury is produced as mercury fulminate.
    ❏ True          ❏ False

72. Mercury and its compounds are not poisonous.
    ❏ True          ❏ False

73. Oxidation symbol is E0.
    ❏ True        ❏ False

74. Pure mercury cannot be mined.
    ❏ True        ❏ False

75. The leading producer of mercury is Spain.
    ❏ True        ❏ False

76. Oxidation is measured in amperes.
    ❏ True        ❏ False

77. Ferric chloride is used to make printed circuit boards.
    ❏ True        ❏ False

78. Potassium has a brown color.
    ❏ True        ❏ False

79. Potassium is so soft it can be cut with a knife.
    ❏ True        ❏ False

80. Lithium is the lightest metal.
    ❏ True        ❏ False

81. Zinc is easily turned to zinc oxide.
    ❏ True        ❏ False

82. Zinc metal is hard and brittle at room temperatures.
    ❏ True        ❏ False

83. Alkali is used to name six chemical elements: lithium, sodium, potassium, rubidium, cesium, and francium.
    ❏ True        ❏ False

84. Morphine and codeine are also known as alkaloids.
    ❏ True        ❏ False

85. Some alkaloids are used as medicines.
    ❏ True        ❏ False

86. Soft solder is defined as the bonding of metals together with a tin alloy that melts under 800°F.
    ❏ True        ❏ False

87. The soldering process is selected by considering the joint design, the product size and the product shape.
    ❏ True        ❏ False

88. The pascal (Pa) has been selected for use in checking pressure; however, in the English system it is expressed as psi (pounds per square inch).
    ❏ True        ❏ False

89. Atmospheric pressure is 14.7 psi.
    ❏ True        ❏ False

90. An atmospheric pressure of 101.3 kPa will balance or support a column of mercury 76 cm high.
    ❏ True        ❏ False

91. The formulas for converting temperatures are $1.8 \times C + 32 = F$ and $(F-32)$ divided by $1.8 = C$.
    ❏ True        ❏ False

92. The value of pi is 3.14159.
    ❏ True        ❏ False

93. The United Kingdom (UK) consists of Scotland, Wales, and England.
    ❏ True        ❏ False

94. The Canadians have converted to the metric system.
    ❏ True        ❏ False

95. The kilopascal is used for fluid pressure for all fields of use in barometric pressure, gas pressure, tire pressure, and water pressure.
    ❏ True        ❏ False

96. Atmospheric pressure of 101.3 kPa will balance or support a column of mercury 100 cm high.
    ❏ True        ❏ False

97. Mass is the resistance to motion.
    ❏ True        ❏ False

98. The basic unit for time is the minute.
    ❏ True              ❏ False

99. A circle represents 360° of movement.
    ❏ True              ❏ False

100. pH is a measure of concentration of hydrogen atoms.
     ❏ True              ❏ False

## Practice Questions: Multiple Choice

1. ANSI is the abbreviation for:
   a. American National School Institute
   b. American Natural Plumbing Standards
   c. American National Standards Institute
   d. American National Plumbing Standards

2. A fitting is a _____ to control and guide the flow of water.
   a. Device
   b. Bucket
   c. Switch
   d. Faucet

3. Examples of fixtures are sinks, bathtubs, and closet bowls.
   a. Fitting
   b. Shower receptors
   c. Draining
   d. Vent

4. GPM stands for:
   a. Flow per minute
   b. Grit per minute
   c. Gross produce matters
   d. Gallons per minute

5. The plumbing industry uses the word _____ in reference to kitchen sinks.
   a. Lavatory
   b. Catcher
   c. Laboratory
   d. Sink

6. Porcelain is used as a coating on _____.
   a. Fixtures
   b. Faucets
   c. Bidets
   d. Pipe

7. What plumbing devices use vitreous china?
   a. Faucets
   b. Pipes
   c. Plumbing fixtures
   d. Plumbing controls

8. A product standard for fittings would specify the maximum amount of _____ material.
   a. Color
   b. Alloy
   c. Ceramic
   d. Paint

9. Materials commonly leached into drinking water distribution systems include _____.
   a. Copper
   b. Lead
   c. Iron
   d. All of the above

10. The Safe Drinking Water Act defines "lead free" as being no more than _____ percent of materials used in solders.
    a. 0.1
    b. 0.5
    c. 0.2
    d. 0.6

11. Potable water is satisfactory for drinking, culinary and _____ purposes.
    a. Domestic
    b. Swimming
    c. Recreational
    d. Fishing

12. A high efficiency toilet allows users to flush
    it with _____ gpf.
    a. 1 gallon
    b. 1.6 gallons
    c. 2 gallons
    d. 5 gallons

13. The most common type of toilet used in
    the U.S. is _____.
    a. Water pressure
    b. Air pressure
    c. Gravity fed
    d. Manually operated

14. In the plumbing industry, low-flow fixtures
    and fittings refer to plumbing products
    that meet the water efficiency standard of
    the Energy Policy Act (EPA) of _____.
    a. 2001
    b. 1906
    c. 1992
    d. 1934

15. A toilet that uses a compressed air device
    to enhance the force of gravity used to
    clean the bowl when the toilet is flushed is
    called _____.
    a. Air flushed toilet
    b. Gravity less toilet
    c. Pressured toilet
    d. Pressure assisted

16. Water Sense helps customers to choose
    water-efficient products by specifying the
    _____ flow rates and performance
    levels.
    a. Low
    b. Maximum
    c. Minimum
    d. High

17. A federal law passed in 1990 prohibits dis-
    crimination against people with _____.
    a. TB
    b. Cancer
    c. Mumps
    d. Disabilities

18. A large and rapid change in water tem-
    perature is defined as _____.
    a. Too hot
    b. Thermal shock
    c. Too cold
    d. Tepid

19. The National Standard Plumbing code is
    used in _____, New Jersey and some
    cities.
    a. Maryland
    b. New York
    c. Alabama
    d. Texas

20. The NSPC (National Association of
    Plumbing, Heating and Cooling) is located
    in _____.
    a. Washington, DC
    b. New York City
    c. Falls Church, VA
    d. Charleston, WV

21. Which plumbing code is used in the
    Western United States?
    a. The New Uniform Plumbing Code
    b. National Plumbing Code
    c. International Plumbing Code
    d. Plumbing Code of America

22. ASTM is updated every _____ years.
    a. 10
    b. 12
    c. 5
    d. 2

23. Where is the International Code Council located?
    a. Dallas, TX
    b. Whittier, CA
    c. Austin, TX
    d. Chicago, IL

24. How does the CIPI Code indicate the latest year for its revision?
    a. By last number after the name
    b. Coded in the name
    c. Updated by showing letter to represent date
    d. By using parentheses

25. Canadian Code specifications ae given in _____.
    a. UK units
    b. Metric
    c. U.S. inches
    d. UK pounds

26. ASTMA-A-74 is a specification for _____.
    a. A rubber gasket
    b. Drain pipes
    c. Hub and spigot cast iron pipe fitting
    d. Vent pipe

27. Bathtubs are available as _____ units or as a combination shower/tub.
    a. Milli
    b. Two
    c. Single
    d. Dual

28. The whirlpool tub is a little more complicated than the regular _____ tubs.
    a. Ceramic
    b. Fiberglass
    c. Cast iron
    d. Wooden

29. The sanitary drainage system relies on _____ for its operation.
    a. Pumps
    b. Fans
    c. Air pressure
    d. Gravity

30. The primary purpose of the drainage system is to _____ fluid waste and organic matter as quickly as possible.
    a. Dispose of
    b. Discharge
    c. Move along
    d. Grind up

31. Cast iron or _____ is used for drainage lines.
    a. Plastic
    b. Brass
    c. Clay
    d. Rubber

32. Pipe is available in straight length sections from 12 to _____ feet long.
    a. 30
    b. 20
    c. 10
    d. 40

33. Brass and copper piping are classified by the same nominal sizes as _____ pipe with two weights each.
    a. Plastic
    b. Iron
    c. Clay
    d. Stainless steel

34. Pipe fittings provide continuity by using couplings, nipples and _____.
    a. Tape
    b. Glue
    c. Solder
    d. Reducers

35. Flow in a piping system is regulated by valves that are specified by type, material, and _____.
   a. Size
   b. Diameter
   c. Length
   d. Thread size

36. Two types of water supply systems are in general use, the down-feed and the _____.
   a. Pumped
   b. Cistern
   c. Rain water
   d. Up feed

37. Cold and hot water supply systems usually rely on copper, _____, wrought iron or galvanized steel pipes.
   a. Brass
   b. Plastic
   c. Clay
   d. Stainless steel

38. Waste water piping in a plumbing system carries all the other wastes except those from the _____.
   a. Bathtub
   b. Dishwasher
   c. Kitchen sink
   d. Toilet

39. Re-vents relieve air pressure on _____ that are vertically downstream from the water closet or toilet.
   a. Fixtures
   b. Couplings
   c. Elbows
   d. Showers

40. The _____ vent is the vent that extends through the roof.
   a. Soil
   b. Bathroom
   c. Sink
   d. Water closet

41. Vents can be classified as wet vents or _____ vents.
   a. Smooth
   b. Door openings
   c. Dry
   d. Air

42. Modern homes use either PVC (plastic) pipe or _____ pipe instead of cast iron.
   a. Copper
   b. Stainless steel
   c. Clay
   d. Brass

43. There are three primary types of pipes used in above ground water systems. They are PEX, copper, and _____.
   a. PVC
   b. CPVC
   c. Brass
   d. PEX

44. PEX is very similar to _____ in appearance.
   a. Polyethylene
   b. Copper
   c. CPVC
   d. Brass

45. PE piping has been around for over 50 years and is liked because of its toughness and _____.
   a. Durability
   b. Color
   c. Strength
   d. Thread ability

46. HDPE pipe is chosen for its ability to be joined by _____.
   a. Glue
   b. Lap molding
   c. Heat fusion
   d. Threads

47. Fusion joining has been used for over
    _____ years.
    a. 10
    b. 30
    c. 20
    d. 40

48. Fittings are available for cast iron pipe,
    drainage, waste and vent type _____
    tubing of all sizes.
    a. Copper
    b. Clay
    c. Brass
    d. Plastic

49. Most pipe has been replaced in modern
    buildings with plastic and _____
    piping.
    a. Galvanized
    b. Clay
    c. Copper
    d. PVC

50. Brass pipe is rarely used because it
    _____.

    a. Is very expensive
    b. Is hard to work with
    c. Only comes in 10-foot lengths
    d. Is damaged easily

51. The PEX system is used because _____.
    a. It is flexible
    b. It can be used in cold or hot water
    c. It is colorful
    d. It is readily available

52. Copper pipe is available in three types, L,
    M, and _____.
    a. A
    b. B
    c. c
    d. K

53. The type of pipe used with hot and cold
    spigots is _____.
    a. Cast iron
    b. Galvanized
    c. Copper
    d. Brass

54. Cast iron has been used in drains and
    _____ for years.
    a. Vents
    b. Spigots
    c. Water lines
    d. Fittings

55. What is used in some jobs and is a much
    newer type of cast iron?
    a. Light weight and hub-less
    b. PEX
    c. Brass
    d. Copper

56. What works well with rubber couplings
    and makes the joint leak proof?
    a. Plastic clamps
    b. Super glue
    c. Metal clamps
    d. Rubber tape

57. Heavy rubber couplings can be used with
    _____.
    a. Galvanized pipe
    b. Cast iron
    c. Copper pipe
    d. Brass pipe

58. Be careful when working with molten lead,
    open fires, and _____ metals.
    a. Cold
    b. Glue
    c. Hot
    d. Tape

59. When working with cast iron pipe, it is best not to use a _____ for cutting.
    a. Chain saw
    b. Hacksaw
    c. Reciprocal saw
    d. Chain cutter

60. A joint runner is used with _____.
    a. Brass pipe
    b. Copper pipe
    c. Galvanized pipe
    d. Cast iron

61. Pouring hot metal is a very common task when working with _____.
    a. Copper pipe
    b. Cast iron pipe
    c. Galvanized pipe
    d. Clay pipe

62. Both _____ and fiber pipes are used for drain or sewer lines outside the home.
    a. Clay pipe
    b. Cast iron pipe
    c. Brass pipe
    d. Copper pipe

63. Probably the most accurate method to cut vitrified clay pipe is to use a _____.
    a. Hand saw
    b. Hatchet
    c. Circular power saw
    d. Cold chisel

64. You can use a _____ to score a mark in vitrified clay pipe.
    a. File
    b. Chalk
    c. Hand saw
    d. Paint

65. When you need to cut fiber pipe to a shorter length, probably the best and easiest way to join them is with a flexible _____ coupling.
    a. Sleeve
    b. Rubber hose
    c. Clamp
    d. Plastic clamp-on

66. Specific gravity is the direct ratio of any liquid's weight to the weight of _____ at 62°F, 62.4 lbs/cu ft. or 8.33 gallons.
    a. Vinegar
    b. Hydrogen chloride
    c. Gasoline
    d. Water

67. Static pressure is water required to fill the _____.
    a. System
    b. Line
    c. Measuring device
    d. Bucket

68. Flow pressure is the pressure the pump must develop to overcome the _____ created by the flow through the system.
    a. Obstructions
    b. Acids
    c. Resistance
    d. Air bubble

69. Viscosity is a property of a _____ that resists any force tending to produce flow.
    a. Water
    b. Acid
    c. Liquid
    d. Steam

70. The minimum recommended clearance around pumps is _____ inches.
    a. 24
    b. 6
    c. 35
    d. 48

71. Specific non-clogging pumps are available to handle _____ sludge.
    a. Liquid
    b. Raw
    c. Solid
    d. Toilet

72. The two types of sewage lift stations are wet well and _____.
    a. Deep well
    b. Shallow well
    c. Dry well
    d. Water well

73. The difference between a sewage pump and a sludge pump is the _____.
    a. Impeller
    b. Case
    c. Motor
    d. Voltage rating

74. The three basic types of pumps are centrifugal, rotary, and _____.
    a. Reciprocal
    b. Liquids
    c. Solids
    d. Sludge

75. The simplex or _____ reciprocating pumps are well suited for requirements in tanneries, sugar refineries and so on.
    a. Low
    b. High
    c. Duplex
    d. Well

76. The general-purpose rotary-type pumps are designed to handle either thick or _____ liquids.
    a. Light
    b. Thin
    c. Acid based
    d. Heavy

77. Centrifugal pumps are also adapted to _____ handling applications.
    a. Water
    b. Solids
    c. Sludge
    d. Light

78. Which pumps are designed to handle organic matter?
    a. Pyrex
    b. Turbine
    c. Magma
    d. Rubber impeller

79. There are four types of toilets or water closet bowls; which doesn't belong here?
    a. Blow-out
    b. Reverse trap
    c. Siphon
    d. None of these

80. The blow-out type of toilet (water closet) is operated with _____ valves only.
    a. Flush
    b. Rounded
    c. Flat
    d. Stop

81. The flushing action of the _____ bowl toilet is started by a jet of water being directed through the leg-up of the trap-way.
    a. Reverse trap
    b. Blow-out
    c. Wash-down
    d. Siphon-jet

82. The toilet fixture located in the bathroom of some upper scale homes is called a _____.
    a. Sitz
    b. Bidet
    c. Sloan valve
    d. Sludge remover

83. This fixture is not usually included in homes.
   a. Toilet
   b. Shower stall
   c. Faucet
   d. Urinal

84. This type of urinal has to be not less than six inches deep with one-piece backs.
   a. Stand up
   b. Automatic flush
   c. Through
   d. Manually operated

85. Most lower consumption urinals call for _____ gallons per flush minimum.
   a. 3.8
   b. 5.2
   c. 1.6
   d. 1.0

86. Low-consumption water toilets and urinals will not function properly on excess _____.
   a. Pressure
   b. Water
   c. Lubricant
   d. Clamp pressure

87. When installing a new regulator from a flex tube diaphragm kit, be sure to push the _____ past the o-ring when installing it.
   a. Regulator
   b. Pipe clamp
   c. Hose
   d. Color bar

88. Every home has a _____.
   a. Basement
   b. Kitchen sink
   c. Dishwasher
   d. Bidet

89. In a kitchen sink installation, the Standard Plumbing Code says it should be located _____ inches from the finished floor.
   a. 31 inches
   b. 45 inches
   c. 22.5 inches
   d. 18 inches

90. Water lines are usually roughed in at _____ from the finished floor.
   a. 4 inches
   b. 10 inches
   c. 20 inches
   d. 6 inches

91. The main copper tube size for short-branch connections can be found _____.
   a. In a chart
   b. On the Internet
   c. In books
   d. In the manufacturer's invoice

92. The American Disabilities Act says the sink should be _____.
   a. Installed so a person sitting in a wheelchair can reach it
   b. 6 inches from the floor level
   c. 52 inches above the floor
   d. Even with the counters

93. Copper tube size is controlled by code. Generally, the mains serving short branch connections are _____.
   a. Up to $\frac{3}{8}$-inch tubing
   b. Up to three branches uses $\frac{1}{2}$-inch tubing
   c. Up to five branches can use $\frac{5}{8}$-inch tubing
   d. Up to ten branches can use 1-inch tubing

94. Pressure available to move water through a distribution system, or a part of it, is the main pressure minus the _____.
   a. Pressure lost in the meter
   b. Pressure lost in the water softener
   c. Pressure lost in highly placed fixture
   d. Pressure lost in the fixtures

**95.** Tube sizes are selected in accordance to the pressure losses caused by _____.
a. Velocity
b. Friction
c. Pressures
d. Heat

**96.** Every sink has a _____ mechanism.
a. Switch
b. Automatic washer
c. Trip lever
d. Manually operated faucet

**97.** Shower pans are required to be made of at least 4-pound sheet _____ or 24 gauge copper with corners folded.
a. Plastic
b. Metal
c. Fiberglass
d. Lead

**98.** A _____ functions in the same manner as a faucet.
a. Diverter
b. Valve
c. Brick
d. Pipe

**99.** Minimum water pressure for a shower head is _____.
a. 30 psi
b. 15 psi
c. 300 kPa
d. 40 psi

**100.** Plastic drain, waste and vent piping has been approved by local and state codes for _____.
a. 10 years
b. many years
c. 20 years
d. 30 years

# C | Answers to Test Questions

# Answers for Chapter Review Questions

## Chapter 1

1. To select a tool that will do the job, make a list of job and tools needed, and then acquire the tools.
2. Depends on the type of job you do; at least 2 with a maximum of 5.
3. The internal wrench is used to hold closet spuds and bath, basin and sink strainers.
4. Wherever you need a wrench to fit a variety of sizes of bolts or nuts.
5. The nipple is a short piece of pipe, usually threaded at both ends.
6. Iron.
7. Sewer systems mainly, or for drainage pipes.
8. A spud wrench is used to hold pipe of various diameters. Usually referred to as a pipe wrench.

## Chapter 2

1. GPM means gallons per minute, GPF mean gallons per flush.
2. A lavatory is the bathroom wash bowl. The word is sometimes used by the general public to mean bathroom.
3. Vitreous china is a compound of ceramic materials fired at high temperatures, then glazed. It is used to give a shiny glow to the toilet, sink, and bidet.
4. To leach means the process of dissolving a solution component out of a constituent material at a wetted surface.
5. Plubum.
6. Proposition 65 deals with California Safe Drinking Water and Toxic Enforcement Act of 1986.
7. Products and buildings are considered "barrier free" if they permit access to all users including those in wheelchairs.

8. Thermal shock is a large and rapid change in water temperature.

## Chapter 3

1. National Standard Plumbing Code is located in Maryland, New Jersey, and some cities.
2. The Uniform Plumbing Code is used in the western United States.
3. American Society for Testing and Materials
4. a. American Society for Testing and Materials
   b. Cast Iron Pipe Institute
   c. International Association of Plumbing and Mechanics officials
   d. International Plumbing Code

## Chapter 4

1. To allow framing a building to leave space for the bathtub.
2. Types of bathtubs: porcelain enameled cast iron tubs, some rectangular, oval, corner, and whirlpool/shower.
3. Usually by shape.
4. It depends on how many people will use it.
5. Mostly the weight of the cast iron tub.

## Chapter 5

1. Copper and PVC pipe or metal and plastic, wet and dry.
2. By using pipe—plastic or metal.
3. There are wet and dry vents—lavatory (sink) usually has a vent. Most sinks can use a vent to operate freely as do toilets and everywhere a sink or bathtub is installed.
4. Usually through the roof.

5. Plastic or copper, or in some cases, plastic coated copper.
6. DWV means drain-waste-vent.
7. Plastic (polyvinyl chloride).

## Chapter 6

1. Copper, PEX (cross-length polyethylene) and CPVC (chlorinated polyvinyl chloride).
2. It can be very difficult to identify; the difference is that the CPVC fittings are larger than PVC.
3. Type M, but type L is preferred.
4. Flexible couplings are used for all types of in-house and sewer connections, drains, waste, vent piping, etc.
5. 40 years.

## Chapter 7

1. Iron.
2. With special cutters, chisel and hammer, hacksaw.
3. Pouring hot lead in the joint to seal it.
4. To allow for movement due to expansion and contraction where temperature varies.
5. Hammer and chisel, hacksaw, chain cutter.
6. A straight piece of pipe.
7. Their weight requires it.

## Chapter 8

1. It usually lifts sewage or sludge.
2. Magna pumps handle crude mixtures in the form of a thin paste.
3. Two types of deep well pumps: (1) jet type and (2) submersible type.
4. Pyrex or glass pumps are used for acids, milk, fruit juices and other acid liquids.
5. Static discharge pumps which is a stationary pump with a static discharge head.
6. Four types of special service pumps: (1) rotary gear pump; (2) self-priming, motor mounted; (3) centrifugal pump; and (4) glass lined pump.
7. Centrifugal pumps are used for sewage and solids-handling applications.

## Chapter 9

1. Schools, public buildings, offices.
2. A reverse trap toilet is used with flush valves or low tanks. It is similar in action to the siphon jet.
3. A jet of water is directed through the up-leg of the trap-way. The trap fills with water and the siphoning action begins.
4. The wash down type of toilet is used where low cost is prime factor.
5. A bidet is usually found in upper scale homes or hospitals. It provides a soothing bath to the lower part of the body after a bowel movement.
6. A sitz bath is used to clean the lower part of the body after a bowel movement. Usually accomplished by the bidet with its ability to wash a warm water stream into the area needing attention. A bidet is usually found in hospitals and upper scale homes.

## Chapter 10

1. A device used in restrooms to allow men to urinate standing up.
2. Most do not allow the in homes.
3. The color code is used to identify the device by its capacity.
4. The regulator controls the water amount per flush.
5. An O-ring is used to prevent leaks. It usually is found around pipes or circular ends of pipe.
6. Helps in the prevention of water leaks by surrounding pipes with a rubber ring.
7. A back-check valve prevents water backflow from the urinal.

8. Usually 1 gallon.
9. Manually and automatically.
10. Usually 1 gallon per flush or 3.785 liters.

## Chapter 11

1. To prepare food for family and to wash dishes.
2. Standard Plumbing Code.
3. Kitchen sinks: 1. Service 2. Single compartment and double compartment.
4. Lavatory means sink—usually at least a kitchen sink and one or more in the bathroom.
5. Pressure losses due to friction: (1) pipe size; (2) pipe length; and (3) faucet.
6. A sink with a rim around the basin, usually part of the sink design.
7. A U-trap has a tendency to eliminate the migration of sewer gas into the house.

## Chapter 12

1. 2.5 gallons per minute/9.4 liters per minute.
2. A diverter causes water to be diverted from the shower feeder pipe to the shower head.
3. Local code refers to shower head: ANSI/ASME5 Standard A 112.18.1M CSA (Canadian) B-125.
4. To prevent mold from forming and various germs from breeding.
5. A shower stall is a small area both with a prescribed base and drain with water contained within.

## Chapter 13

1. It comes in 8 to 10 foot lengths bundled in wire or plastic.
2. Threading makes the pipe ends weak.
3. Rigid copper pipe is usually soldered together. Fittings are usually soldered to the pipe.

4. CPVC stands for chlorinated polyvinyl chloride.
5. Zinc.
6. On pipe threads.
7. Galvanized pipe will last indefinitely.
8. Remove burrs from pipe for easier soldering or threading.
9. Usually metal to metal or adapters for plastic pipe to metal pipe. Crimp type.
10. NIBCO fittings represent a drainage system with adapters.

## Chapter 14

1. To cause the solids to liquify.
2. A drain field is a series of ditches properly filled and covered as prescribed by local laws.
3. A cistern is usually dug by specialized contractors, then the walls of the hole are coated with concrete.
4. To allow the bacteria time to do its job and liquify the solids.
5. Professional well diggers or drillers.
6. Water collection system: cistern receives rain run-off from the house as it is directed through eves, troughs and down-spouts to the cistern.
7. Concrete. Some are made of fiberglass.
8. Usually sand is employed as a filter in municipal systems.
9. It is trucked in.
10. From municipal water systems.

## Chapter 15

1. 6-inch plastic pipe.
2. Bituminous pipe is very weak or fragile, and no longer in use. Asbestos pipe is restricted to exterior by local codes. Some places will not allow its use at all.
3. Load factor depends on the load placed on the buried pipe once it is covered (like heavy loads of lumber or a new building).

4. VCP, PVC
5. PVC, VCP
6. It can resist acids.
7. Vitrium is Latin for glass.
8. In laboratories, in chemical plants, and anywhere acids are handled.
9. Clay containing sand
10. HUD government housing for the low-income.

## Chapter 16

1. Municipal systems
2. EPA
3. Drinking
4. Legally
5. MUDs
6. Usually in the terrace (the strip of land between the sidewalk and street curb) near the house
7. Near the street main water line.
8. Many places
9. Environmental Protection Agency
10. Washington, DC

## Chapter 17

1. In drafting rooms where buildings are designed and in books for plumbers and engineers, as well as architects.
2. Symbol for lavatory
3. Wall-hung urinal symbol
4. Kitchen sink symbol
5. American Standard Graphical Symbols
6. In mathematics.
7. Square root sign ($\sqrt{\phantom{x}}$)
8. Cu is sign for copper.
9. $2 \times 2 = 4$, or $2^2$
10. $C^2 = \sqrt{A^2 + B^2}$

## Chapter 18

1. Gold
2. It forms a thin film on the surface when it oxidizes and this prevents rusting. Zinc oxide is produced by this oxidation process.
3. An alkaloid is a group of six chemicals that react violently in water. Some are used in medicine. Nicotine is an alkaloid.
4. Soft solder uses lead and zinc. It bonds metal under 800°F.
5. Solder used to hold copper pipes together and prevent leaks.
6. Zinc is mined and is always found in combination with sulfur or it's called zinc sphalterite.
7. In soft solder.
8. Metal not found in hard form is potassium.
9. Oxidation potential is a measure of an element's tendency to oxidize or lose electrons.
10. Hg.

## Chapter 19

1. UK and European continent.
2. Pascal (Pa)
3. It's ¼ of a cord of wood. A stere is 1 cubic meter.
4. Pi = 3.14159, usually rounded to 3.1416.
5. The meter is the unit of measure in the metric system.
6. 1 decimeter (dm) = 0.1 of a meter = 3.957 inches.
7. Pressure.
8. HP uses Polish Notation, others use Scientific Notation.

## Chapter 20

1. The metric system is used in Europe
2. The English System of measure is used in the U.S.
3. Conversion means to change from one system of measure to another.
4. UK is made up of England, Wales, and Scotland.
5. The 1970s.

## Chapter 21

1. One thousandth of a meter
2. Pascal (Pa)
3. Use a formula or chart.
4. England and most of Europe, as well as Canada.
5. 110° is 43.3°C
6. Celsius
7. 72°F is 22.2°C
8. −10°F is −23.3°C
9. 10°C is 50°F
10. 304.8 mm to U.S. or UK measurement units is 0.3048 meter or about 12 inches.

## Chapter 22

1. Meter
2. Mole measures the amount of substance.
3. Light
4. Second
5. Water boils at 212°F or 100°C
6. 360 degrees
7. 4
8. Meter
9. 3.1459
10. 3

## Chapter 23

1. Sulfuric acid
2. $H_2SO_4$
3. In an apple
4. Salicylic acid
5. Hydrochloric acid is used in gold mining.
6. In the home, office, schools
7. Alkali is opposite of acid.
8. Acids are inorganic.
9. pH stands for the hydrogen ion that has a positive charge.
10. pH of water is the measurement with litmus paper that turns red when the hydrogen is present; when it turns blue it means the solution has lost its hydrogen ions.

## Chapter 24

1. Loading and unloading trucks or trains when located in noisy areas.
2. To communicate with another on the job when there is a noisy background.
3. (1) Bowline; (2) running bowline; (3) rolling hitch; (4) catspaw; and (5) square knot.
4. Both ends of the rope have loops that tighten up when pulled and hold like they originally did.
5. When working with large concrete pipe and heavy steel pipe.

## Chapter 25

1. To protect the public from unsafe drinking water and unsafe waste disposal facilities.
2. One who is beginning to learn a trade.
3. Usually four years.
4. Get a sponsor plumber and fill out the required forms.
5. Attend a trade school or work for four years with a master plumber and take the required exam.
6. Be an experienced licensed plumber and pass the state's test.

7. Reciprocity does not exist for journeyman and contractor plumbers.

8. Texas and Nevada

9. ADA affects plumbers. They have to attend classes on barrier-free provisions and Uniform Construction Code requirements.

10. A back flow device is usually found on hose connections where the hose attaches to the house for watering trees, plants, and lawn.

# Answers for Practice Questions: True/False

| | | | |
|---|---|---|---|
| 1. T | 26. T | 51. T | 76. T |
| 2. T | 27. T | 52. T | 77. T |
| 3. F | 28. T | 53. F | 78. F |
| 4. F | 29. F | 54. T | 79. T |
| 5. T | 30. T | 55. T | 80. T |
| 6. T | 31. T | 56. T | 81. T |
| 7. F | 32. T | 57. F | 82. T |
| 8. T | 33. F | 58. T | 83. T |
| 9. T | 34. F | 59. T | 84. T |
| 10. T | 35. T | 60. T | 85. T |
| 11. F | 36. T | 61. T | 86. F |
| 12. T | 37. T | 62. T | 87. T |
| 13. T | 38. T | 63. T | 88. T |
| 14. T | 39. F | 64. T | 89. T |
| 1S. F | 40. T | 65. T | 90. T |
| 16. F | 41. T | 66. T | 91. T |
| 17. T | 42. F | 67. T | 92. T |
| 18. T | 43. F | 68. T | 93. T |
| 19. T | 44. T | 69. T | 94. T |
| 20. T | 45. T | 70. T | 95. T |
| 21. T | 46. T | 71. T | 96. F |
| 22. F | 47. T | 72. F | 97. T |
| 23. T | 48. F | 73. T | 98. F |
| 24. F | 49. F | 74. F | 99. T |
| 25. F | 50. T | 75. T | 100. T |

# Answers for Practice Questions: Multiple Choice

| | | | |
|---|---|---|---|
| 1. c | 26. c | 51. b | 76. b |
| 2. a | 27. c | 52. d | 77. b |
| 3. b | 28. c | 53. a | 78. c |
| 4. d | 29. d | 54. a | 79. c |
| 5. d | 30. a | 55. a | 80. a |
| 6. a | 31. a | 56. c | 81. d |
| 7. c | 32. b | 57. b | 82. b |
| 8. b | 33. b | 58. c | 83. d |
| 9. d | 34. d | 59. b | 84. c |
| 10. c | 35. a | 60. d | 85. c |
| 11. a | 36. d | 61. b | 86. b |
| 12. b | 37. d | 62. a | 87. a |
| 13. c | 38. d | 63. c | 88. b |
| 14. c | 39. a | 64. a | 89. c |
| 15. d | 40. a | 65. a | 90. d |
| 16. c | 41. c | 66. d | 91. a |
| 17. d | 42. a | 67. a | 92. a |
| 18. b | 43 b | 68. c | 93. a |
| 19. a | 44. a | 69. c | 94. a |
| 20. c | 45. a | 70. a | 95. b |
| 21. a | 46. c | 71. b | 96. c |
| 22. c | 47. d | 72. c | 97. d |
| 23. b | 48. a | 73. d | 98. b |
| 24. a | 49. c | 74. a | 99. b |
| 25. b | 50. a | 75. c | 100. b |

# D | Plumbing Specifications

# Specifications for Plumbing Buildings

A.  Provide all materials and equipment and perform all labor required to install, complete and operate plumbing systems as indicated on the drawings, and as specified and required by the local code.

B.  Run all soil, waste and vent piping with 2 percent minimum grade unless otherwise noted (edit the slope to suit project requirements). Horizontal vent piping shall be graded to drip back to the soil or waste pipe by gravity.

C.  Elevations as shown on the drawings are to the bottom of all pressure piping and to the invert of all gravity piping.

D.  Adjust sewer inverts to keep the tops of pipes in line where the pipe's size changes.

E.  Maintain a minimum of 3'6" of ground cover over all underground water mains and a minimum of 3'0" of ground cover over all underground sewers and drains. Edit the depth of the ground cover to suit frost line depth and project requirements.

F.  Provide shutoff valves in all domestic water piping system branches in which branch piping serves two or more fixtures.

G.  Unless otherwise noted, all domestic cold- and hot water piping shall be of the ½-inch size. Edit the system type or pipe size to suit project requirements.

H.  Unless otherwise noted, all piping is overhead, tight to the underside of the slab, with space for insulation if required.

I.  Install piping so all valves, strainers, unions, traps, flanges, and other appurtenances requiring access are accessible.

J.  Where domestic cold and hot water piping drops into a pipe chase, the size shown for the pipe drops shall be used to the last fixture.

K.  Install all piping without forcing or springing.

L.  All piping shall clear doors and windows.

M.  All piping shall grade to low points. Provide hose end drain valves at the bottom of all risers and low points.

N.  Unions and/or flanges shall be installed at each piece of equipment, in bypasses, and in long piping runs (100 ft. or more) to permit disassembly for alteration and repairs.

O.  All valves shall be adjusted for smooth and easy operation.

P.  All valves (except control valves) and strainers shall be the full size of the pipe before reducing the size to make connections to the equipment and controls.

Q.  Provide chain-wheel operators for all valves in equipment roams mounted greater than 7'0" above floor level; chain shall extend to 7'0" above floor level.

R.  Provide all plumbing fixtures and equipment with accessible stops.

S.  Unless otherwise noted, drains shall be installed at the low point of roofs, areaways, floors, and so forth.

T.  Provide cleanouts in sanitary and storm drainage systems at end of runs, at changes in direction, near the base of stacks, every 50 ft. in horizontal runs and elsewhere as indicated. Edit horizontal cleanout spacing to suit code and project requirements.

U. All cleanouts shall be the full size of the pipe for pipe sizes 6 in. and smaller and shall be 6 in. for pipe sizes larger than 6 in.

V. All balancing valves and butterfly valves shall be provided with position indicators and maximum adjustable stops (memory stops).

W. All valves shall be installed so the valve remains in service when the equipment or piping on the equipment side of the valve is removed.

X. All piping work shall be coordinated with all trades involved. Offsets in piping around obstructions shall be provided at no additional cost to the owner.

Y. Provide flexible connections in all piping systems connected to pumps and other equipment that require vibration isolation. Flexible connections shall be provided as close to the equipment as possible or as indicated on the drawings.

# Americans with Disabilities Act (ADA) Requirements

A. ADA Titles—American Disabilities Act
1. Title I: Equal Employment Opportunity
2. Title II: State and Local Governments
3. Title III: New and Existing Public Accommodations and New Commercial Facilities
4. Title IV: Telecommunications
5. Title V: Miscellaneous Provisions

B. Drinking Fountains
1. Where only one drinking fountain is provided on a floor, a drinking fountain with two bowls, one high bowl and one low bowl, is required.
2. Where more than one drinking fountain is provided on a floor, 50 percent shall be handicapped accessible and shall be on an accessible route.
3. Spouts shall be no higher than 36 in. above the finished floor or grade.
4. Spouts shall be located at the front of the unit and shall direct the water flow parallel or nearly parallel to the front of the unit.
5. Controls shall be mounted on the front or side of the unit.
6. Clearances:
   a. should be 27 in. high, 30 in. wide, and 17–19 in. deep, with a minimum front clear floor space of 30 in. × 48 in.
   b. Units without clear space below: 30 in. × 48 in. Clearance is suitable for parallel approach.

C. Water Closets
1. The height of the water closet shall be 17–19 in. to the top of the toilet seat.
2. Flush controls shall be hand-operated or automatic. Controls shall be mounted on the wide side of toilet areas, and no more than 44 in. above the floor.
3. At least one toilet shall be handicapped accessible.

D. Urinals
1. Urinals shall be stall-type or wall hung with an elongated rim at a maximum of 17 in. above the floor.
2. Flush controls shall be hand operated or automatic. Controls shall be mounted no more than 44 in. above the floor.
3. If urinals are provided, at least one shall be handicapped accessible.

E.   Lavatories
    1.   Lavatories shall be mounted with the rim or counter surface no higher than 34 in. above the finished floor with a clearance of at least 29 in. to the bottom of the apron.
    2.   Hot water and drain pipe under lavatories shall be insulated or otherwise configured to protect against contact.
    3.   Faucets shall be lever-operated, push-type, and electronically controlled. Self-closing valves are acceptable, provided they remain open a minimum of 10 seconds.

F.   Bathtubs
    1.   Bathtub controls shall be located toward the front half of the bathtub.
    2.   Shower units shall be provided with a hose at least 60 in. long that can be used both as a fixed shower head and a handheld shower head.

G.   Shower Stalls
    1.   The shower controls shall be opposite the seat in a 36 in. × 36 in. shower stall and adjacent to the seat in a 30 in. × 60 in. shower stall.
    2.   Shower units shall be provided with a hose at least 60 in. long that can be used both as a fixed shower head and a handheld shower head.

H.   Forward Reach
    1.   Maximum high forward reach: 48 in.
    2.   Side Reach
    3.   Maximum high side reach: 54 in.
    4.   Minimum low side reach: 9 in.

# Index

*References to figures are in italics.*